The War That's Already Started

A Sobering Examination of Global Conflict

By Dave Denness

Published by Self-Published via Kindle Direct Publishing Printed in Australia

Complete Edition, 2025

ISBN: 978-1-7642050-0-9

Dedication

This book draws deep inspiration from the Diary of a CEO (DOAC) podcast episode titled: WW3 Threat Assessment: The West is collapsing; can we stop it?! They want you confused and obedient! In this 2025 DOAC roundtable, host Steven Bartlett brings together three renowned experts: former CIA intelligence officer Andrew Bustamante, nuclear war journalist Annie Jacobsen, and global politics expert Benjamin Radd.

Together, they dissect the most pressing threats facing humanity, from nuclear escalation to AI-driven manipulation, amassing over 2.5 million views at the time of writing. Their stark warning resonates: the West teeters on collapse amid rising nuclear dangers, with tensions boiling between the US and Iran, Russia and Ukraine, and China and Taiwan. Are we barreling toward World War 3, especially under a Trump administration?

These voices from intelligence, journalism, and analysis sound an urgent alarm. As a dedicated podcast listener, I make it a point to seek balanced perspectives by exploring a variety of sources and opinions. DOAC stands out for its raw, unfiltered approach where Steven Bartlett interviews trailblazers in thought, business, and culture to reveal profound truths about success, setbacks, mental health, and growth. It represents equal parts self-improvement and compelling narrative, challenging norms and sparking real transformation. (And no, I am not compensated for this endorsement.)

Table of Contents

Introduction: Why This Matters

It was a quiet evening when I started to watch the podcast that inspired this book. I have always been interested in espionage and the hidden forces shaping global events through media manipulation. What unfolded was a chilling examination of a world already at war—not with conventional armies or aerial bombardments, but through invisible digital strikes that could destabilise everything we hold dear.

As the experts discussed cyber infiltration of power grids, hypersonic missiles, and disinformation campaigns designed to fracture societies from within, I felt a profound shift in my understanding. This was not some distant future scenario; it was our current reality, a hidden conflict influencing every digital interaction, every news story, and every decision we make. This book tells that story: the war that has already begun, woven into the fabric of our interconnected world, and how we might still change its trajectory.

I am writing this for you: the parent worried about your children's future, the professional scrolling through news feeds, the citizen sensing the world is slipping off balance. No jargon or academic walls here; just clear stories drawn from real events and expert insights to show how cyber breaches can empty your bank account overnight, nuclear decisions hang on a leader's frayed nerves, and fake narratives spark real chaos in your community. We will walk through it together, like unravelling a thriller where the heroes are us, armed with knowledge and action.

Picture a hacker in a dimly lit room, fingers flying over keys to shut down a nation's lights. Or a missile streaking across the sky so fast there is no time to react. Or a viral post that turns neighbors into enemies. These are not movie plots; they are headlines from 2024, like the Colonial Pipeline hack that left Americans scrambling for gas, or Iran's missiles lighting up Israel's night sky. Ignored, they lead to darkness, collapsed economies, nuclear winters, lost freedoms. But seen clearly, they become opportunities to build back stronger.

We will start in the shadows of the digital battlefield, then face the cold stare of nuclear reality, navigate the high-stakes games nations play, uncover the mind-bending information wars, and end with your toolkit for resilience. This represents a path from fear to empowerment, because this war hits home: cyber spikes up 30% last year, tensions in Ukraine and the Middle East boiling over, China's moves in the South China Sea pushing buttons.

How do we live well in this storm? That is the heart of our story. Let us begin with the conversation that opened my eyes, and yours too, I hope.

I have maintained the structure from the original podcast that inspired this book as it flowed well.

Part I: Recognising the Battlefield

Chapter 1: The Conversation That Changed Everything

I remember the moment vividly: sitting in my living room as the podcast hosts explored a discussion that fundamentally challenged my understanding of modern conflict. War, they explained, no longer resembled our traditional images of grand armies marching across battlefields or fighter jets filling the skies.

Instead, contemporary warfare operates like a digital spectre, infiltrating systems through cyber vulnerabilities to destabilise our interconnected world. The experts described hackers quietly commandeering power grids, sophisticated algorithms delivering targeted disinformation to divide communities, and economic systems manipulated with surgical precision. It was as if a veil had been lifted, revealing a conflict already in progress—one that had infiltrated our daily lives without our awareness.

That realisation sparked this book, a quest to understand and share how we can stand against it.

In this new era, battles are not confined to fields; they are part of a vast, interconnected web where physical force is just one strand. Cyber operations have become the go-to method for striking without boots on the ground. Take the 2020 SolarWinds cyberattack: Russian-affiliated hackers compromised Orion software updates, creating backdoor access to approximately 18,000 organisations, including multiple US government agencies and Fortune 500 companies. Rather than an isolated incident, this represented a systematic campaign to establish persistent access to critical infrastructure and sensitive information systems.

Beyond direct cyber-attacks on infrastructure, modern warfare employs disinformation as a strategic weapon—a silent saboteur that can destabilise entire societies without conventional explosives. The 2016 US election interference demonstrated this approach, with carefully crafted falsehoods designed to exploit existing social divisions and undermine democratic institutions. These represent not crude fabrications but sophisticated psychological operations, amplified by technology to reach hearts and minds in seconds.

This transformation of warfare has democratised conflict participation. Today's battlefields include not only traditional military personnel but also civilian programmers launching cyber operations, content creators developing propaganda narratives, and ordinary citizens who unknowingly amplify disinformation through social media engagement. The front lines now encompass servers guarding our secrets, applications molding our views, and markets steering our futures. Seeing this represents our first weapon.

The consequences hit hard and fast. When Russian-affiliated hackers crippled the Colonial Pipeline in 2021, fuel lines dried up, sparking chaos at pumps and a taste of what a full-scale digital siege could bring—blacked-out cities, empty shelves, lives on hold.

What makes it terrifying is the stealth: no smoke rising, no ruins to mourn. A grid could be rigged for months, waiting for the signal. Disinformation simmers, quietly corroding unity until the pot boils over.

This conflict extends beyond state actors; underdogs with gadgets can punch above their weight. A solo hacker or fringe group can crash systems or swing elections, turning the world into a wild card game.

To endure, we redefine defense: it is not just for armies, but for all of us: locking data, questioning sources, building bonds. This chapter maps the terrain; next, the digital depths.

The cast is wide: superpowers like the US, China, Russia, but also Iran, North Korea, terrorists wielding the same arsenal. No solo hero wins this.

Economics weaves in: sanctions aimed at regimes crush everyday folks, inflating prices and sparking unrest. This economic dimension of modern warfare leaves innocents in the lurch. The podcast peeled back the human layer: our digital addiction leaves doors ajar; every interaction creates a vulnerability. But our responses matter: fact-checking, wise trust, and genuine connections hold the key. Disinformation preys on feelings; a rage- bait post goes viral, like the 2020 US unrest fanned by outsiders exploiting biases.

The current global situation demonstrates these principles in action. Russia's re-invasion of Ukraine and Israel's wars with Palestine and Iran show the devastation of drones and bombs on civilian infrastructure and the obliteration of whole cities. However, the war that has already started does not look like the wars we have seen before. It is being fought firstly in cyberspace, in the global economy, in the media, and in the minds of ordinary citizens who do not even realise they are participants in a global conflict.

Foreign intelligence services are already conducting operations inside the United States, not just gathering information, but actively working to undermine our democratic institutions and turn Americans against each other. Cyber-attacks on critical infrastructure, disinformation campaigns designed to sow confusion and discord, and economic warfare that is designed to weaken our ability to respond to military threats.

The nuclear landscape has changed dramatically in just the past few years, with new weapons systems that compress the timeline for decision-making and increase the risk of miscalculation. The traditional deterrence structures that maintained stability during the Cold War are breaking down in an era of multiple nuclear powers and increasingly sophisticated delivery systems. The global political order is shifting in ways that make conflict more likely, not less. The rise of authoritarian and megalomaniac regimes, the breakdown of international institutions, and the increasing competition for resources and influence is driving major powers toward confrontation.

This represents a war that is being fought on multiple fronts, using weapons and tactics that most people do not understand, with stakes that could not be higher, even if most of us do not recognise it yet.

Now, you might be thinking, "Okay, this is all very interesting, but what does it have to do with my life? I am not a spy or a nuclear weapons expert or a global politics analyst. I am just trying to pay my bills and take care of my family." That is exactly why this conversation matters so much.

Because the war that has already started is not just affecting governments and military leaders—it is affecting all of us, in ways that we are only beginning to understand as commodity prices around the world increase whilst the average workers' wages have remained stagnant for years. Cyber-attacks are not just targeting government computers—they are targeting the power grid that keeps your lights on, the financial systems that protect your savings, and the communication networks that connect you to the world.

Nuclear weapons are not just threatening other countries—they are threatening every major city in the world, including the one where you live. The global political trends are not just academic exercises—they are determining whether your children will grow up in a world of peace and prosperity or conflict and scarcity.

The threats we are facing today are real, they are immediate, and they are affecting all of us, whether we realise it or not. The question is not whether these threats exist—the question is whether we are going to understand them well enough to respond effectively. World War III is not inevitable, and nuclear war is not just around the corner; however, we are at a critical moment in history, a moment when the decisions we make as individuals and as a society will determine what kind of world we leave for future generations. But we can only make good decisions if we understand what we are dealing with. We can only respond effectively to these threats if we know they exist and understand how they work. And we can only protect ourselves and our families if we are willing to look beyond the comfortable illusions that most of us prefer to the uncomfortable realities that these experts deal with every day.

That is what this book is about—helping you understand the world as it really is, not as we wish it were. It is about taking insights from these experts and making them accessible to everyone who wants to understand what is really happening and what they can do about it.

This started with a simple question about whether we are already at war. By the time you finish this book, you will not only understand the answer to that question—you will understand what it means for your life and what you can do to navigate the challenges ahead. Because the war has already started.

The question is: are you ready to understand what that means for you?

Chapter 2: Digital Warfare and the New Battleground

Gone are the days of war as spectacle—armies clashing, skies thundering. Now, it represents a quiet invasion in the glow of screens, where data streams become rivers of attack, algorithms stir storms of division, and the networks knitting our world turn into traps. We scroll and click oblivious, until a glitch hits home: no power, frozen funds, or a lie that sparks fury. Our digital habits, news feeds, chats, transactions—are steps on mined ground. The net, once a wonder, is the epicenter of ceaseless strife.

Remember when war meant armies facing off on battlefields, tanks rolling across borders, and fighter jets screaming through the sky? Those traditional images of conflict seem almost quaint compared to the reality of modern warfare. The war that is happening right now is being fought on servers, through counter intelligence services, media algorithms and in the digital infrastructure that keeps our entire civilisation running. And here is really concerning part—most of us do not even know we are under attack unless you personally are affected by hacking.

We are not talking about phishing or African princes promising you a share of their fortune; this is far more complex and organised. Think about your typical day. You wake up and check your phone for news. You scroll through social media to see what your friends are up to. You check your bank account, maybe do some online shopping, send some emails for work. You are living your life, minding your own business, completely unaware that every single one of those digital interactions is taking place on a battlefield. That is not hyperbole—that is the reality of modern warfare. The internet is not just a tool for communication and commerce anymore. It is the primary theatre of operations for a global conflict that is literally happening 24 hours a day, 7 days a week, 365 days a year.

Imagine if during World War II, the Germans had been able to walk into American factories, American banks, American power plants, and American government buildings whenever they wanted, without anyone noticing. That is essentially what is happening now, except instead of walking through doors, they are walking through fibre optic cables. The difference is that in World War II, if enemy soldiers showed up in your town, you would know about it. You would see them, you would hear the gunfire, you would know you were under attack. But when foreign hackers infiltrate your power grid or your financial services systems or your social media feeds, you do not see anything. Life goes on as normal, right up until the moment when the lights go out or your bank account gets emptied, or you find yourself believing things that are not true.

When most people think about cyber-attacks, they picture some teenager in a hoodie trying to steal credit card numbers. But the cyber warfare that governments deal with today is something entirely different. We are talking about nation-state actors with unlimited budgets and teams of the world's best hackers, targeting the most critical systems that keep our society functioning.

In 2021, a group called DarkSide—which intelligence experts believe has connections to the Russian government—shut down the Colonial Pipeline, which supplies gasoline to much of the Eastern

United States. For six days, gas stations ran dry, people panicked, and the entire region got a taste of what happens when critical infrastructure gets targeted by cyber-attacks. But here is what really should scare you—that attack was relatively minor compared to what is possible. The hackers who shut down the Colonial Pipeline were essentially just trying to make money through ransomware. They were not trying to cause maximum damage or chaos. They were just criminals looking for a payday. Now imagine what would happen if a foreign government, or worse still AI or rogue intelligence, decided to launch a coordinated cyber-attack designed not to make money, but to cause maximum disruption to society.

Imagine if they simultaneously targeted the power grid, the financial services systems, the telecommunications networks and the transportation infrastructure. Imagine if they did it during a natural disaster or a political crisis, when our ability to respond would be compromised. A nuclear weapon can destroy a city, but a coordinated cyber-attack could destroy a nation, and the really terrifying part is that cyber-attacks are much easier to launch than nuclear weapons, much harder to trace back to their source, and much more difficult to defend against.

Let us talk about something that most of us take completely for granted—electricity. You flip a switch, the lights come on. You plug in your phone, it charges. You turn on your air conditioning, your house gets cool. It is so reliable that we barely think about it, which is exactly why it is such an attractive target for our enemies.

The power grid is a marvel of engineering, but it is also incredibly vulnerable to cyber-attacks. Most of the systems that control our power plants and electrical distribution networks were designed decades ago, long before anyone was thinking about cyber security. They were built to be efficient and reliable, not to withstand attacks from sophisticated hackers.

Foreign intelligence services have already infiltrated many of these systems. They are not necessarily doing anything destructive right now—they are just positioning themselves so that they could cause massive damage if they ever decided to. It is like having enemy soldiers hiding in your basement, waiting for orders to attack.

The scary thing about this is that we know they are there. Our intelligence agencies have detected foreign hackers in our power grid systems, but getting them out is incredibly difficult, and keeping them out is even harder. It is like trying to secure a building that has thousands of doors and windows, and you can only watch a few of them at a time. The implications of a major attack on the power grid are almost too scary to think about. No electricity means no refrigeration, so food spoils. No electricity means no electricity for hospitals, no electricity for water treatment plants, no electricity for banks and financial systems. In a matter of hours, a modern society could be thrown back decades in terms of basic functionality.

The financial system represents another critical vulnerability that most people take for granted until it fails. Every time you use a credit card, check your bank balance online, or make an electronic payment, you are relying on a complex network of interconnected computer systems that are constantly under attack. Foreign hackers have already penetrated many of these systems, not necessarily to steal money immediately, but to understand how they work and to position themselves for future attacks.

In 2016, hackers linked to North Korea successfully stole $81 million from the Bangladesh Bank by exploiting vulnerabilities in the SWIFT international banking system. This was not a sophisticated technical attack; it was largely successful because of social engineering and insider knowledge of banking procedures. The scary part is that this attack could have been much worse. The hackers initially tried to steal nearly $1 billion, and they were only stopped by a typo in one of their transfer requests that raised suspicions.

Now imagine what would happen if a foreign government decided to launch a coordinated attack on the global financial system during a time of international crisis. They could potentially freeze bank accounts, manipulate currency exchange rates, disrupt stock markets, and create financial panic that could trigger a global economic collapse. The interconnected nature of modern financial systems means that an attack on one major bank or financial institution could have cascading effects throughout the entire global economy.

Chapter 3: The Forgotten Wars - Sudan, Myanmar, and Haiti

While the world's attention focuses on the high-profile conflicts in Ukraine and the Middle East, three devastating wars rage in different corners of the globe, each demonstrating unique aspects of modern conflict and state failure. These forgotten wars in Sudan, Myanmar, and Haiti reveal how quickly political crises can spiral into humanitarian catastrophes, how traditional concepts of state sovereignty are being challenged, and how criminal organisations can effectively replace government authority.

Sudan: When Political Transition Becomes Civil War

The Sudan Civil War, which erupted in April 2023, represents one of the most devastating conflicts currently unfolding worldwide, yet it receives minimal international attention compared to other global crises. The conflict began as a power struggle between the Sudanese Armed Forces (SAF) and the Rapid Support Forces (RSF), two military factions that had previously cooperated in overthrowing the civilian government but turned against each other in a brutal fight for control.

The speed and intensity of Sudan's descent into chaos demonstrates how rapidly modern conflicts can escalate beyond anyone's control. Between April 15, 2023, and October 25, 2024, the warring sides exchanged a total of 8,942 attacks, averaging sixteen attacks per day. This represents one of the most intense conflict environments globally, with fighting spreading from the capital Khartoum to multiple regions across the country.

The humanitarian consequences have been catastrophic. More than 25 million people—over half of Sudan's population—now require humanitarian assistance, with over 14 million of these being children. The conflict has displaced more than one million people internally, while hundreds of thousands have fled to neighbouring countries, creating a regional refugee crisis that strains the resources of Chad, Egypt, Ethiopia, and other neighbouring states.

In August 2024, the Famine Review Committee officially confirmed famine conditions in the Darfur region of Sudan, marking the first time famine has been officially declared anywhere in the world since 2017. This famine affects some of the most vulnerable populations in one of the world's poorest countries, with children particularly at risk of malnutrition and death.

The conflict's regional implications extend far beyond Sudan's borders. The breakdown of state authority has created opportunities for arms trafficking, human trafficking, and the spread of extremist groups. Neighbouring countries face increasing pressure from refugee flows, while the disruption of agricultural production in Sudan—traditionally a significant food producer for the region—contributes to food insecurity across East Africa.

What makes Sudan's conflict particularly relevant for understanding modern warfare is how it demonstrates the fragility of political transitions in the contemporary international environment. Sudan had been attempting to transition from military rule to civilian democracy following the 2019 overthrow of longtime dictator Omar al-Bashir. However, the military coup in October 2021 derailed this transition, and the subsequent power struggle between military factions has destroyed any hope of peaceful democratic development.

The international community's response to Sudan's crisis has been largely ineffective, highlighting the limitations of traditional diplomatic and humanitarian tools in addressing complex internal conflicts. Despite multiple ceasefire agreements and international mediation efforts, the fighting has continued to intensify, with both sides committing serious violations of international humanitarian law.

Myanmar: The Collapse of Military Authority

The Myanmar Civil War, ongoing since the military coup of February 1, 2021, represents a fundamental challenge to traditional concepts of state sovereignty and territorial control. What began as protests against military rule has evolved into a nationwide armed resistance that has effectively fragmented the country and reduced the military government's control to just 21 per cent of Myanmar's territory.

The scope and scale of the resistance movement in Myanmar is unprecedented in modern conflict. More than 2,600 rebel groups are currently fighting the military junta, ranging from established ethnic armed organisations that have been in conflict with the central government for decades to newly formed People's Defense Forces (PDF) composed of civilians who took up arms after the coup.

The human cost of Myanmar's civil war has been enormous. More than 82,000 people have been killed since the coup began, including over 6,000 civilians. Between 3.2 and 3.4 million people have been internally displaced, while approximately 149,000 have fled to neighboring countries. The military has been systematically bombing civilian sites, including schools, hospitals, and religious buildings, in what human rights organisations describe as a deliberate campaign to terrorize the population into submission.

What makes Myanmar's conflict particularly significant for understanding modern warfare is how it demonstrates the role of technology and social media in organising resistance movements. Despite the military's attempts to control internet access and communications, opposition groups have used encrypted messaging apps, social media platforms, and digital currencies to coordinate activities, share information, and raise funds for their operations.

The involvement of ethnic armed groups in training and supporting the civilian resistance has created a unique dynamic in Myanmar's conflict. These groups, which have been fighting the central government for decades over issues of autonomy and self-determination, have found common cause with the pro-democracy movement, creating a broader coalition against military rule than has existed in previous conflicts.

The international response to Myanmar's crisis has been complicated by the country's strategic location between China and India, two major powers with competing interests in the region. China has maintained relations with the military government while also engaging with some ethnic armed groups along its border. India has been concerned about refugee flows and the potential for instability to spread across its border with Myanmar.

The Association of Southeast Asian Nations (ASEAN), of which Myanmar is a member, has struggled to respond effectively to the crisis due to its principle of non-interference in member states' internal affairs. This has highlighted the limitations of regional organisations in addressing serious human rights violations and armed conflicts within their membership.

Haiti: When Criminal Organizations Replace the State

The Haiti crisis represents a unique form of modern conflict where criminal organisations have effectively replaced state authority across large portions of the country. The situation reached catastrophic levels in 2024, with 5,601 murders recorded—an increase of 1,000 deaths compared to 2023— establishing a national homicide rate of nearly 48 per 100,000 people.

The scale of criminal control in Haiti is unprecedented in the Western Hemisphere. An estimated 200 to 300 armed gangs are currently active throughout the country, with criminal groups united under the "Viv Ansanm" (Living Together) coalition controlling approximately 60 per cent of Port-au-Prince, the capital city. These gangs have moved beyond traditional criminal activities to establish territorial control, collect taxes, and provide basic services in areas where the state has completely withdrawn.

The March 2024 jailbreak that resulted in over 4,700 inmates escaping from Haiti's two largest prisons demonstrated the complete breakdown of state authority and the gangs' ability to conduct coordinated operations against government institutions. This mass prison break was not simply a criminal act, but a strategic operation designed to recruit additional fighters and demonstrate the gangs' power relative to the state.

The humanitarian consequences of Haiti's crisis have been severe. More than one million people have been internally displaced, with many living in overcrowded camps with limited access to food, water, and medical care.

The collapse of basic services, including healthcare, education, and sanitation, has created conditions that facilitate the spread of disease and increase mortality rates, particularly among children and elderly populations.

The international response to Haiti's crisis has included the deployment of a Kenyan-led Multinational Security Support (MSS) mission, authorised by the UN Security Council in October 2023. However, this mission has faced significant challenges, including limited resources, unclear mandates, and the complexity of operating in an environment where the distinction between state and non-state actors has largely disappeared.

What makes Haiti's crisis particularly relevant for understanding modern conflict is how it demonstrates the potential for state collapse in the Western Hemisphere and its direct implications for U.S. security interests. The breakdown of authority in Haiti has created opportunities for drug trafficking, human trafficking, and other transnational criminal activities that directly affect American communities.

The crisis has also generated significant migration pressures, with thousands of Haitians attempting to reach the United States and other countries in the region. This created political challenges for the Biden administration at that time highlighted the interconnected nature of security challenges in the Americas.

Lessons from the Forgotten Wars

These three conflicts—Sudan, Myanmar, and Haiti—illustrate several important trends in contemporary warfare that readers need to understand. First, they demonstrate how quickly political crises can escalate into full-scale conflicts with regional and international implications. In each case, what began as internal political disputes rapidly evolved into complex humanitarian emergencies that overwhelmed local capacity to respond.

Second, these conflicts show how traditional concepts of state sovereignty and territorial control are being challenged in the modern era. In Myanmar, the military government controls only a fraction of the country's territory. In Haiti, criminal organisations have effectively replaced state authority in large areas. In Sudan, competing military factions have divided the country between them, with neither able to establish legitimate governance.

Third, these conflicts highlight the limitations of international institutions and traditional diplomatic tools in addressing complex internal conflicts. Despite extensive international attention and resources, the international community has been largely unable to prevent or resolve any of these crises, raising questions about the effectiveness of current approaches to conflict prevention and resolution.

Finally, these forgotten wars demonstrate the interconnected nature of modern security challenges. Each conflict has regional implications that affect neighboring countries and global implications that affect international markets, migration patterns, and security threats.

Understanding these connections is essential for comprehending how local conflicts can have global consequences in an interconnected world.

: The AI Revolution in Warfare

of artificial intelligence into military systems
the most significant developments in warfare since
nuclear weapons. Unlike previous military technologies that enhanced existing capabilities, AI has the potential to fundamentally change the nature of conflict by accelerating decision-making processes, reducing human oversight, and creating new categories of weapons that can operate independently of direct human control.

Current AI Military Applications

Artificial intelligence is already being integrated into military systems across multiple domains, from intelligence analysis to logistics optimization to weapons targeting. The United States Department of Defense has identified AI as a critical technology for maintaining military superiority, investing billions of dollars annually in AI research and development. China has similarly prioritised military AI development as part of its broader strategy to challenge U.S. military dominance.

Current AI applications in military systems include automated target recognition, which uses machine learning algorithms to identify and classify potential targets from satellite imagery, drone footage, and other surveillance data. These systems can process vast amounts of visual information far more quickly than human analysts, enabling rapid identification of threats and opportunities on the battlefield.

Predictive maintenance systems use AI to analyse data from military equipment to predict when components are likely to fail, enabling proactive maintenance that reduces downtime and improves operational readiness.

Logistics optimisation systems use AI to manage complex supply chains, optimising the movement of personnel, equipment, and supplies across global military operations.

Intelligence analysis has been revolutionised by AI systems that can process and analyse vast amounts of data from multiple sources, identifying patterns and connections that human analysts might miss. These systems can monitor social media, communications intercepts, financial transactions, and other data sources to provide early warning of potential threats and opportunities.

Autonomous Weapons Development

The development of lethal autonomous weapons systems (LAWS) represents one of the most controversial applications of AI in warfare. These systems, sometimes called "killer robots" by critics, are designed to select and engage targets without direct human intervention. While fully autonomous weapons may not yet exist in operational military forces, the technology to create them is rapidly advancing.

Current semi-autonomous weapons systems, such as the Israeli Iron Dome missile defense system and various naval close-in weapons systems, can engage targets automatically once activated by human operators. However, these systems operate within carefully defined parameters and against specific types of threats. The next generation of autonomous weapons would have much broader capabilities and greater independence from human control.

The debate over autonomous weapons centers on fundamental questions about human control over lethal force and the ethical implications of allowing machines to make life-and-death decisions. Proponents argue that autonomous weapons could reduce civilian casualties by making more precise targeting decisions and could protect military personnel by operating in dangerous environments without risking human lives.

Critics argue that autonomous weapons would lower the threshold for armed conflict by removing human emotional and psychological barriers to killing. These autonomous weapons could malfunction or be hacked by adversaries and would violate fundamental principles of human dignity and the laws of war that require human judgement in targeting decisions.

The international community has struggled to develop governance frameworks for autonomous weapons. The Convention on Certain Conventional Weapons has held multiple meetings to discuss potential regulations but has not reached consensus on binding restrictions. Some countries, including the United States and Russia, have opposed comprehensive bans on autonomous weapons, while others have called for preemptive prohibition.

AI-Driven Cyber Warfare

Artificial intelligence is revolutionizing cyber warfare by enabling more sophisticated and adaptive attacks that can evolve in real-time to overcome defensive measures. Traditional cyber-attacks rely on pre-programmed exploits and attack vectors that can be detected and countered by security systems. AI-driven attacks can adapt their behavior based on the responses of target systems, making them much more difficult to defend against.

Machine learning algorithms can be used to identify vulnerabilities in target systems by analysing network traffic, system configurations, and user behavior patterns. These algorithms can discover previously unknown vulnerabilities and develop exploits automatically, significantly reducing the time and expertise required to conduct cyber-attacks.

AI can also be used to conduct social engineering attacks at scale, using natural language processing to generate convincing phishing emails, social media posts, and other communications designed to trick users into revealing sensitive information or installing malware. These AI-generated communications can be personalised for individual targets based on their social media profiles, online behavior, and other available data.

Defensive applications of AI in cyber warfare include automated threat detection systems that can identify and respond to attacks in real-time, behavioral analysis systems that can detect unusual network activity that might indicate an ongoing attack, and automated response systems that can isolate compromised systems and deploy countermeasures without human intervention.

The arms race between AI-powered cyber-attacks and AI-powered cyber defenses is accelerating rapidly, with both offensive and defensive capabilities advancing simultaneously. This creates a dynamic environment where the advantage can shift quickly between attackers and defenders, making it difficult to predict the long-term implications for cyber security.

Information Warfare Enhancement

Artificial intelligence has dramatically enhanced the capabilities of information warfare by enabling the creation of convincing false content at unprecedented scale and sophistication. Deepfake technology, which uses AI to create realistic but fabricated audio and video content, has the potential to undermine trust in authentic information and create confusion about what is real and what is artificial.

AI-generated text can be used to create fake news articles, social media posts, and other written content that is difficult to distinguish from human-created content. These systems can generate content in multiple languages and can be trained to mimic the writing styles of specific individuals or organisations, making the false content more convincing to target audiences.

Social media manipulation has been enhanced by AI systems that can create and manage large networks of fake accounts, automatically generate content designed to influence public opinion, and identify and exploit social divisions within target populations. These systems can operate at scale across multiple platforms simultaneously, making them much more effective than traditional propaganda techniques.

AI can also be used to analyse the effectiveness of information warfare campaigns in real-time, adjusting messaging and targeting based on audience responses and engagement metrics. This enables more sophisticated and adaptive information operations that can evolve based on their success in achieving desired outcomes.

The implications of AI-enhanced information warfare extend far beyond military conflicts to include election interference, social manipulation, and the erosion of shared factual foundations for democratic discourse. The ability to create convincing false content at scale threatens to undermine the information environment that democratic societies depend on for informed decision-making.

Ethical and Legal Challenges

The integration of AI into warfare raises fundamental ethical and legal questions that the international community is struggling to address.

Traditional laws of war, including the Geneva Conventions and their Additional Protocols, were developed for human-controlled weapons and may not adequately address the unique challenges posed by AI-powered systems.

The principle of distinction, which requires parties to a conflict to distinguish between combatants and civilians, becomes more complex when applied to AI systems that must make these determinations automatically.

The principle of proportionality, which requires that the expected civilian harm from an attack not be excessive compared to the anticipated military advantage, requires human judgement that may be difficult to program into AI systems.

Accountability for the actions of AI weapons systems presents another significant challenge. If an autonomous weapon commits a war crime or causes unintended civilian casualties, it is unclear who should be held responsible: the programmer who created the system, the commander who deployed it, the operator who activated it, or the political leader who authorised its use.

The speed of AI decision-making also raises concerns about human oversight and control. If AI systems can make targeting decisions in milliseconds, there may not be sufficient time for meaningful human review and approval, effectively removing human control over life-and-death decisions even in systems that are nominally under human command.

Future Implications

The future implications of AI in warfare are profound and far-reaching. As AI systems become more sophisticated and autonomous, they may fundamentally change the nature of military conflict by accelerating the pace of operations beyond human comprehension and control.

AI-powered military systems may create new forms of deterrence and escalation dynamics that are not yet fully understood. The speed and unpredictability of AI decision-making could increase the risk of accidental conflicts or rapid escalation of minor incidents into major confrontations.

The proliferation of AI military technologies to non-state actors and smaller countries could democratise advanced military capabilities, potentially levelling the playing field between major powers and smaller adversaries.

This could make conflicts more unpredictable and difficult to control.

The integration of AI into nuclear command and control systems raises particularly serious concerns about the risk of accidental nuclear war. If AI systems are given authority over nuclear weapons, technical malfunctions or adversarial manipulation could potentially trigger nuclear conflicts without human intention or authorisation.

The development of AI warfare capabilities is likely to drive an arms race among major powers, with each seeking to maintain or achieve superiority in this critical technology domain. This competition could destabilise international relations and increase the risk of conflict as countries seek to test and demonstrate their AI military capabilities.

Understanding these implications is essential for citizens, policymakers, and military leaders as they navigate the challenges and opportunities presented by the AI revolution in warfare. The decisions made today about the development, deployment, and governance of AI military systems will shape the nature of conflict for generations to come.

Part II: The Nuclear Reality

Chapter 5: Speed Kills: The Hypersonic Threat

The development of hypersonic weapons represents one of the most significant advances in military technology since the advent of nuclear weapons themselves. These weapons, capable of travelling at speeds exceeding Mach 5 (five times the speed of sound), fundamentally alter the strategic calculus of nuclear deterrence by compressing decision-making timelines to mere minutes and rendering traditional missile defense systems obsolete.

The physics of hypersonic flight creates unique challenges for both offensive and defensive systems. Unlike ballistic missiles, which follow predictable parabolic trajectories through space, hypersonic weapons can maneuver throughout their flight path while maintaining extreme speeds. This maneuverability, combined with their velocity, makes them virtually impossible to intercept with current defensive technologies.

Russia has taken the lead in hypersonic weapons development, deploying the Kinzhal air-launched hypersonic missile and the Avangard hypersonic glide vehicle. China has developed the DF-ZF hypersonic glide vehicle and has conducted numerous tests of hypersonic systems. The United States, despite its technological advantages in many military domains, has lagged behind in hypersonic weapons development, though it is now investing heavily in catching up.

The strategic implications of hypersonic weapons extend far beyond their immediate military applications. The compressed timeline for decision-making that these weapons create increases the risk of miscalculation and accidental escalation. When leaders have only minutes to decide whether an incoming hypersonic weapon carries a conventional or nuclear warhead, the pressure to respond quickly may override careful analysis and diplomatic communication.

The deployment of hypersonic weapons also undermines the stability that has characterised nuclear deterrence for decades. The traditional model of deterrence relies on the ability to detect incoming threats, assess their nature, and respond appropriately. Hypersonic weapons compress this timeline so dramatically that meaningful human decision-making becomes extremely difficult, if not impossible.

Perhaps most concerning is the potential for hypersonic weapons to be integrated with artificial intelligence systems that could make targeting and engagement decisions without human oversight. The speed of hypersonic weapons may necessitate automated response systems that can react faster than human operators, effectively removing human control from critical decisions about the use of force.

The proliferation of hypersonic technology to additional countries and potentially non-state actors represents another significant concern. As the technology becomes more accessible, the number of actors capable of launching devastating attacks with minimal warning will increase, creating a more complex and unpredictable threat environment.

Current missile defense systems, including the Ground-based Midcourse Defense system in the United States and similar systems in other countries, were designed to intercept ballistic missiles during their predictable flight paths through space. Hypersonic weapons, with their ability to maneuver at low altitudes throughout their flight, render these systems largely ineffective.

The development of new defensive technologies specifically designed to counter hypersonic threats is underway, but these systems face enormous technical challenges. The speed and maneuverability of hypersonic weapons require defensive systems that can detect, track, and intercept targets moving at unprecedented velocities while changing course unpredictably.

The economic implications of the hypersonic arms race are also significant. Developing effective hypersonic weapons and the defensive systems needed to counter them requires massive investments in research, development, and deployment. This creates pressure on military budgets and may divert resources from other critical defense priorities.

The psychological impact of hypersonic weapons on both military planners and civilian populations cannot be underestimated. The knowledge that devastating attacks can be launched with virtually no warning creates a climate of constant vulnerability that may influence decision-making in ways that increase rather than decrease the risk of conflict.

Understanding the hypersonic threat is essential for comprehending the current state of global security. These weapons represent a fundamental shift in the nature of warfare and deterrence, creating new vulnerabilities and instabilities that will shape international relations for decades to come.

Chapter 6: Inside the Nuclear Launch Process

The process by which nuclear weapons are authorised, launched, and delivered represents one of the most tightly controlled and carefully designed systems in human history. Understanding this process is crucial for comprehending both the safeguards that exist to prevent accidental nuclear war and the vulnerabilities that could lead to catastrophic miscalculation.

In the United States, the nuclear command and control system is designed around the principle of ensuring that the President, as Commander-in-Chief, maintains ultimate authority over nuclear weapons while also providing the capability to respond rapidly to threats. The famous "nuclear football"—a briefcase containing communication equipment and authentication codes— accompanies the President at all times, symbolising this awesome responsibility.

The process begins with threat detection through a network of satellites, radar systems, and other sensors designed to identify incoming missiles or other nuclear threats. This information is immediately relayed to the North American Aerospace Defense Command (NORAD) and the Pentagon, where analysts must quickly determine the nature and scope of the threat.

If an incoming threat is confirmed, the President is immediately notified and presented with a range of response options. These options, contained in the Single Integrated Operational Plan (SIOP), include everything from limited nuclear strikes against specific targets to full-scale retaliation against an adversary's entire nuclear arsenal and military infrastructure.

The President has approximately six to twelve minutes to make this decision—the time it takes for an intercontinental ballistic missile to travel from Russia or China to the United States. This compressed timeline leaves little room for deliberation, consultation, or verification of the threat. The system is designed for speed, not careful analysis.

Once the President authorises a nuclear response, the order is transmitted through the National Military Command Center to the relevant military commanders. These commanders must verify the authenticity of the order using special authentication codes before proceeding with the launch sequence.

The actual launch process varies depending on the type of nuclear weapon being deployed. For land-based intercontinental ballistic missiles, the order is transmitted to missile silos where two-person crews must simultaneously turn keys to initiate the launch sequence. For submarine-launched ballistic missiles, similar procedures exist aboard nuclear submarines, which may be operating independently in remote ocean locations.

The human element in this process represents both a safeguard and a potential vulnerability. While human oversight is intended to prevent accidental launches, the extreme time pressure and stress of a nuclear crisis could lead to errors in judgement or communication failures that result in unintended escalation.

The integration of artificial intelligence and automated systems into nuclear command and control raises new questions about human control over these ultimate weapons. While AI systems could potentially improve the speed and accuracy of threat assessment, they also introduce new possibilities for technical malfunctions or adversarial manipulation that could trigger nuclear responses without human intention.

Other nuclear powers have developed their own command and control systems, each with different procedures and safeguards. Russia's system, known as "Perimeter" or "Dead Hand," is designed to ensure retaliation even if the country's leadership is eliminated in a first strike. China's nuclear command structure is less well understood but is believed to maintain tight centralised control over nuclear weapons.

The proliferation of nuclear weapons to additional countries has created multiple independent nuclear command systems, each with its own procedures, safeguards, and vulnerabilities. This multiplicity increases the complexity of nuclear crisis management and the potential for miscalculation or technical failure.

Recent technological developments have introduced new challenges to nuclear command and control. Cyber-attacks against command systems could potentially disrupt communications, provide false information about threats, or even interfere with launch procedures. The increasing sophistication of cyber warfare capabilities makes these threats more credible and concerning.

The psychological pressure on decision-makers in nuclear crises cannot be underestimated. The knowledge that millions of lives hang in the balance, combined with extreme time pressure and incomplete information, creates conditions where even experienced leaders may make decisions they would not make under normal circumstances.

Training and simulation exercises are conducted regularly to prepare military personnel and political leaders for nuclear crisis scenarios.

However, these exercises cannot fully replicate the psychological stress and uncertainty of an actual nuclear crisis, leaving questions about how well-prepared decision-makers truly are for such situations.

The nuclear launch process represents humanity's most dangerous gamble—a system designed to prevent nuclear war through the credible threat of nuclear retaliation, but which also creates the possibility of accidental or unintended nuclear conflict. Understanding this system is essential for comprehending the nuclear risks that continue to threaten human civilisation.

Chapter 7: Surviving the Nuclear Threat

The prospect of nuclear conflict, while hopefully remote, requires serious consideration of survival strategies and preparedness measures.

Understanding how to respond to nuclear threats and attacks could mean the difference between life and death for millions of people. This chapter examines both individual and societal approaches to nuclear survival, from immediate response to long-term recovery.

The immediate effects of a nuclear explosion are devastating and occur within seconds to minutes of detonation. The initial nuclear radiation, intense heat, and massive blast wave can cause immediate death and destruction within several kilometers of ground zero. However, survival is possible even relatively close to a nuclear explosion if appropriate protective measures are taken quickly.

The most critical factor in surviving a nuclear attack is understanding the concept of time, distance, and shielding. Time refers to minimising exposure duration to radiation; distance means getting as far as possible from the source of radiation; and shielding involves placing dense materials between yourself and the radiation source.

In the event of a nuclear explosion, the first priority is to seek immediate shelter. If you are outdoors when a nuclear weapon detonates, you should immediately lie flat on the ground and cover your head. The blast wave will arrive within seconds, followed by intense heat and radiation. Do not look at the flash, as it can cause permanent blindness.

If you are indoors when a nuclear weapon detonates, move to the center of the building, away from windows and exterior walls. Basements provide the best protection, particularly if they are below ground level. The goal is to put as much dense material as possible between yourself and the outside environment.

The fallout period following a nuclear explosion presents ongoing dangers that can persist for days or weeks. Radioactive particles carried by wind can contaminate large areas downwind from the explosion site. Understanding wind patterns and fallout predictions can help determine the safest areas and the appropriate duration of shelter.

The "7:10 Rule" provides a useful guideline for radiation decay: for every seven-fold increase in time after detonation, there is a ten-fold decrease in radiation levels. This means that radiation levels one hour after detonation will be ten times lower after seven hours, and one hundred times lower after forty-nine hours.

Emergency supplies are crucial for surviving the fallout period. A basic nuclear survival kit should include water (one gallon per person per day for at least two weeks), non-perishable food, battery-powered or hand-crank radio, flashlights, first aid supplies, medications, dust masks, plastic   sheeting and duct tape for sealing rooms, and sanitation supplies.

Communication during a nuclear emergency may be severely disrupted. Cell phone networks, internet services, and electrical power may be damaged or overloaded. Battery-powered or hand-crank radios become essential for receiving emergency information and instructions from authorities.

Decontamination procedures are important for removing radioactive particles from clothing, skin, and hair. If you have been exposed to fallout, remove outer clothing and shoes before entering shelter, wash exposed skin with soap and water, and avoid scrubbing, which can drive radioactive particles deeper into the skin.

Medical considerations in nuclear survival include understanding radiation sickness symptoms and treatment options. Potassium iodide tablets can help protect the thyroid gland from radioactive iodine, but they must be taken before or shortly after exposure to be effective. These tablets do not provide protection against other types of radiation.

Community preparedness significantly improves survival chances for entire populations. Communities should develop evacuation plans, identify suitable shelter locations, establish communication protocols, and maintain emergency supply stockpiles. Regular drills and education programs help ensure that residents know how to respond effectively.

Government emergency response systems play a crucial role in nuclear survival. Emergency Alert Systems can provide immediate warnings and instructions, while emergency management agencies coordinate evacuation, shelter, and recovery operations. Understanding how these systems work and how to access emergency information is essential for effective response.

The psychological aspects of nuclear survival are often overlooked but critically important. The stress and trauma of a nuclear attack can impair decision-making and reduce survival chances. Mental health support, stress management techniques, and maintaining hope and purpose are essential components of nuclear survival.

Long-term recovery from nuclear attacks involves environmental cleanup, infrastructure rebuilding, and addressing ongoing health effects.

Radioactive contamination can persist for years or decades, requiring careful monitoring and remediation efforts. Understanding these long-term challenges helps in planning for post-attack recovery.

International cooperation is essential for effective nuclear emergency response. Nuclear accidents or attacks can have effects that cross national borders, requiring coordinated international response efforts. International organisations like the International Atomic Energy Agency provide frameworks for cooperation and assistance.

Prevention remains the most effective approach to nuclear survival. Supporting nuclear disarmament efforts, strengthening international institutions, and promoting diplomatic solutions to conflicts all contribute to reducing nuclear risks. Individual actions, such as staying informed about nuclear issues and supporting responsible policies, can contribute to prevention efforts.

Education and awareness are crucial components of nuclear preparedness. Understanding nuclear threats, survival techniques, and emergency procedures empowers individuals and communities to respond effectively. Regular training and updates ensure that knowledge remains current and actionable.

While the prospect of nuclear conflict is frightening, knowledge and preparation can significantly improve survival chances. By understanding the immediate and long-term effects of nuclear weapons, developing appropriate emergency plans, and maintaining necessary supplies, individuals and communities can increase their resilience in the face of nuclear threats.

The goal is not to create fear or panic, but to provide practical information that could save lives in the event of a nuclear emergency. Preparation for nuclear threats is similar to preparation for other natural disasters—it requires planning, supplies, and knowledge, but it is entirely achievable for individuals and communities willing to take the necessary steps.

Part III: The Global Chess Game

Chapter 8: Israel, America, and the Proxy Game

The relationship between Israel and the United States represents one of the most significant and complex alliances in contemporary international relations, with profound implications for Middle Eastern stability and global security. This partnership, forged through decades of shared strategic interests, common values, and mutual security concerns, has become a central element in the broader geopolitical competition between major powers in one of the world's most volatile regions.

The foundation of the U.S.-Israel relationship rests on multiple pillars: strategic cooperation, intelligence sharing, military assistance, and diplomatic support. The United States provides Israel with approximately$3.8 billion in military aid annually, making it the largest recipient of U.S. foreign assistance. This aid package, formalised in a ten-year Memorandum of Understanding signed in 2016, ensures Israel's qualitative military edge in the region while also supporting American defense industries.

Intelligence cooperation between the two countries has been particularly close, with both nations benefiting from shared information about regional threats, terrorism, and weapons proliferation. Israeli intelligence services have provided valuable insights into Middle Eastern affairs, while American intelligence capabilities have enhanced Israel's understanding of global threats and strategic developments.

The proxy dimension of Middle Eastern conflicts has become

increasingly prominent in recent years, with regional powers using allied groups and organisations to advance their interests while avoiding direct confrontation. Iran has been particularly effective in this approach, supporting proxy forces across the region including Hezbollah in Lebanon, Hamas in Gaza, the Houthis in Yemen, and various Shia militias in Iraq and Syria.

Israel's security challenges are fundamentally shaped by this proxy warfare environment. The country faces threats not only from state actors but also from non-state groups that receive funding, training, and weapons from hostile regional powers. This creates a complex security environment where traditional deterrence strategies may be less effective and where conflicts can escalate rapidly across multiple fronts.

The October 7, 2023, Hamas attack on Israel and the subsequent conflict in Gaza demonstrated the devastating potential of proxy warfare in the modern era. The attack, which killed over 1,200 Israelis and resulted in the taking of hundreds of hostages, represented one of the most significant intelligence failures in Israeli history and triggered a massive military response that has had regional and global implications.

The Gaza conflict has highlighted the challenges of fighting proxy forces that operate from densely populated civilian areas and use civilian infrastructure for military purposes. The high civilian casualty toll in Gaza has created international pressure on Israel while also demonstrating the complex ethical and legal challenges of modern warfare in urban environments.

Iran's role as the primary sponsor of anti-Israeli proxy forces has made it a central focus of Israeli strategic planning. The Islamic Republic's nuclear program, its support for terrorist organisations, and its efforts to establish a "ring of fire" around Israel through proxy forces have created an existential threat that Israeli leaders believe requires active countermeasures.

The Abraham Accords, signed in 2020, represented a significant shift in Middle Eastern geopolitics by normalising relations between Israel and several Arab states, including the United Arab Emirates, Bahrain, Morocco, and Sudan. These agreements were facilitated by shared concerns about Iranian regional ambitions and demonstrated the potential for new alliance structures in the Middle East.

However, the Gaza conflict has complicated efforts to expand the Abraham Accords, particularly regarding potential normalisation between Israel and Saudi Arabia. The Saudi leadership has indicated that progress on Palestinian statehood would be necessary for any normalisation agreement, creating a complex diplomatic challenge for all parties involved.

The United States finds itself in a delicate position as it seeks to balance its support for Israel with broader regional relationships and global strategic interests. American military involvement in the region has included direct support for Israel's defense against Iranian missile attacks, while also attempting to prevent regional escalation that could draw the United States into a broader conflict.

The proxy warfare model has implications that extend far beyond the Middle East. The success or failure of proxy strategies in the region influences similar approaches in other parts of the world, including Ukraine, where Russia has used proxy forces, and the

South China Sea, where China employs maritime militias to advance its territorial claims.

Economic dimensions of the proxy conflict include the use of sanctions, trade restrictions, and financial warfare to pressure adversaries and support allies. The United States has imposed extensive sanctions on Iran and its proxy networks, while also providing economic support to allies and partners in the region.

The information warfare component of Middle Eastern conflicts has become increasingly sophisticated, with all parties using social media, traditional media, and diplomatic channels to shape international opinion and justify their actions. The battle for narrative control has become as important as military operations in determining the outcomes of conflicts.

Regional powers beyond Iran and Israel also play significant roles in the proxy warfare environment. Turkey, Saudi Arabia, Egypt, and other countries support various groups and factions based on their own strategic interests, creating a complex web of competing alliances and rivalries.

The humanitarian consequences of proxy warfare in the Middle East have been severe, with civilian populations bearing the brunt of conflicts they did not choose. The displacement of millions of people, the destruction of infrastructure, and the breakdown of basic services have created humanitarian crises that require international attention and assistance.

Looking forward, the proxy warfare model in the Middle East is likely to continue evolving as regional powers adapt their strategies to changing circumstances. The integration of new technologies, including artificial intelligence, cyber capabilities, and advanced weapons systems, will create new opportunities and challenges for all parties involved.

The U.S.-Israel relationship will continue to be tested by these evolving challenges, requiring careful calibration of support levels, strategic coordination, and diplomatic engagement. The ability of both countries to adapt their approaches to the changing nature of regional threats will be crucial for maintaining stability and advancing their shared interests.

Understanding the dynamics of proxy warfare in the Middle East is essential for comprehending broader patterns of international conflict and competition. The lessons learned from this region's experiences with proxy warfare will likely influence strategic thinking and policy development in other parts of the world where similar dynamics are emerging.

Chapter 9: Economic Warfare in the 21st Century

Economic warfare has emerged as one of the most powerful and frequently used tools of international competition in the modern era. Unlike traditional military conflict, economic warfare can inflict significant damage on adversaries while maintaining plausible deniability and avoiding the immediate escalation risks associated with kinetic operations. This form of conflict operates through sanctions, trade restrictions, financial system manipulation, and technological competition, creating new battlefields in global markets and supply chains.

The weaponisation of economic interdependence represents a fundamental shift in how nations compete and conflict with one another. The same global economic integration that has driven prosperity and development for decades has also created vulnerabilities that can be exploited by adversaries seeking to advance their strategic interests without resorting to traditional warfare.

Sanctions have become the primary tool of economic warfare, with major powers using them to pressure adversaries, punish undesirable behavior, and signal resolve to allies and partners. The United States, in particular, has leveraged its dominant position in the global financial system to impose sanctions that can effectively cut targeted countries and entities off from international commerce.

The sanctions regime against Russia following its re-invasion of Ukraine represents one of the most comprehensive economic warfare campaigns in modern history. These measures have targeted Russian banks, energy exports, technology imports, and individual oligarchs and officials. The goal has been to degrade Russia's ability to finance its military operations while imposing costs that might influence its strategic calculations.

However, the effectiveness of sanctions as a tool of economic warfare remains debated. While sanctions can impose significant economic costs, they often fail to achieve their stated political objectives and may have unintended consequences that harm innocent populations or strengthen authoritarian control. The sanctions against Russia, for example, have contributed to global inflation and energy shortages while having uncertain effects on Russian decision-making.

China's economic rise has created new dimensions of economic warfare, with the United States and its allies increasingly concerned about Chinese economic practices that they view as unfair or threatening. Trade wars, technology restrictions, and investment screening mechanisms have become tools for managing economic competition with China while protecting national security interests.

The concept of economic decoupling or "de-risking" has gained prominence as countries seek to reduce their dependence on potentially hostile economic partners. This process involves reshoring critical manufacturing, diversifying supply chains, and restricting technology transfers that could enhance adversaries' military capabilities.

Supply chain warfare has emerged as a particularly important dimension of economic conflict. The COVID-19 pandemic demonstrated the vulnerability of global supply chains to disruption, while also highlighting the strategic importance of controlling key nodes in these networks. Countries are now viewing supply chain resilience as a national security priority.

The semiconductor industry has become a central battleground in economic warfare, with the United States imposing extensive restrictions on Chinese access to advanced chip technology. These measures are designed to prevent China from developing military capabilities that could threaten American interests while maintaining U.S. technological superiority in critical areas.

Financial warfare represents another significant dimension of economic conflict. The ability to exclude adversaries from international payment systems, freeze assets, and manipulate currency markets provides powerful tools for economic coercion. The SWIFT international banking system has become a weapon in this context, with the exclusion of Russian banks demonstrating the power of financial infrastructure control.

Cyber-attacks on economic targets have become increasingly common, with state and non-state actors targeting financial institutions, energy infrastructure, and commercial enterprises. These attacks can cause immediate economic damage while also undermining confidence in digital systems that underpin modern commerce.

Energy has long been a tool of economic warfare, with suppliers using their market position to influence the behavior of importing countries. Russia's use of energy exports as a geopolitical weapon in Europe has been a prominent example, leading to efforts by

European countries to diversify their energy sources and reduce dependence on Russian supplies.

The weaponisation of technology standards and platforms has created new forms of economic warfare. Control over critical technologies, from 5G networks to artificial intelligence systems, provides leverage over countries that depend on these technologies for their economic development and national security.

Economic espionage and intellectual property theft represent covert forms of economic warfare that can provide significant advantages to perpetrating countries. The theft of trade secrets, research data, and proprietary technologies can save years of development time and billions of dollars in research costs.

The role of private companies in economic warfare has become increasingly important, with governments using regulatory powers, market access, and other tools to influence corporate behavior in ways that advance national interests. Companies find themselves caught between competing national demands and must navigate complex geopolitical considerations in their business decisions.

International institutions and agreements have become battlegrounds for economic warfare, with countries seeking to shape rules and norms in ways that advantage their economic interests. Trade organisations, standard-setting bodies, and multilateral agreements all become venues for economic competition and conflict.

The humanitarian consequences of economic warfare can be severe, particularly for civilian populations in targeted countries. Sanctions and trade restrictions can limit access to essential goods, medical supplies, and technologies needed for economic development, creating suffering among populations that may have little influence over their governments' policies.

Defensive measures against economic warfare include diversifying economic relationships, building strategic reserves, developing domestic capabilities in critical sectors, and creating alternative institutions and systems that reduce dependence on potentially hostile partners.

The future of economic warfare will likely be shaped by technological developments, changing global economic structures, and evolving geopolitical relationships. Artificial intelligence, quantum computing, biotechnology, and other emerging technologies will create new opportunities for economic competition and conflict.

Understanding economic warfare is essential for businesses, policymakers, and citizens who must navigate an increasingly complex global economic environment. The integration of economic and security considerations requires new approaches to policy-making and strategic planning that account for the weaponisation of economic relationships.

The challenge for democratic societies is to develop effective tools for economic competition while maintaining open economies and respect for international law. This requires careful balance between security concerns and economic efficiency, as well as coordination among allies and partners to maximize the effectiveness of economic measures.

Economic warfare represents a fundamental challenge to the liberal international economic order that has underpinned global prosperity for decades. How countries respond to this challenge will shape the future of international economic relations and determine whether economic integration continues to be a source of mutual benefit or becomes primarily a tool of conflict and coercion.

Chapter 10: China's Long Game and the South China Sea

China's rise as a global superpower represents one of the most significant geopolitical developments of the 21st century, with profound implications for international stability, economic relationships, and the future of the global order. Nowhere is this more evident than in the South China Sea, where China's assertive territorial claims and military buildup have created one of the world's most dangerous potential flashpoints for great power conflict.

The South China Sea encompasses approximately 3.5 million square kilometers of ocean, through which roughly one-third of global maritime trade passes annually. This includes energy shipments, manufactured goods, and raw materials that are essential to the global economy. Control over these sea lanes provides enormous strategic and economic leverage, making the region a natural focus for great power competition.

China's claims in the South China Sea are based on the "nine-dash line," a boundary that encompasses roughly 90% of the sea and overlaps with the territorial claims of Vietnam, the Philippines, Malaysia, Brunei, and Taiwan. These claims have no basis in international law, as confirmed by a 2016 ruling by the Permanent Court of Arbitration in The Hague, which China has refused to recognise or implement.

The artificial island construction campaign that China has undertaken since 2013 represents one of the most significant military engineering projects in modern history. China has built seven artificial islands on previously submerged reefs and rocks, adding over 3,200 acres of new land to disputed features. These islands have been equipped with military-grade airstrips, radar systems, missile batteries, and other military infrastructure.

The strategic implications of China's island-building campaign extend far beyond the immediate region. These installations provide China with the ability to project military power throughout the South China Sea, potentially denying access to U.S. and allied forces in the event of a conflict. The islands also serve as forward bases for intelligence collection and surveillance of military activities in the region.

China's approach to the South China Sea reflects a broader strategic concept known as "salami slicing"—the gradual accumulation of small changes that individually seem insignificant but collectively result in major strategic shifts. This approach allows China to advance its interests while avoiding the kind of dramatic escalation that might provoke a strong international response.

The use of "grey zone" tactics has been central to China's South China Sea strategy. These tactics involve the use of civilian vessels, coast guard ships, and maritime militias to assert territorial claims and harass other countries' activities without crossing the threshold that would trigger a military response. This approach creates ambiguity about the nature of Chinese activities and makes it difficult for other countries to respond appropriately.

The People's Liberation Army Navy (PLAN) has undergone dramatic modernisation and expansion over the past two decades, transforming from a coastal defense force into a blue-water navy capable of operating globally. This naval buildup has been accompanied by the development of advanced anti-ship missiles, submarines, and other systems designed to challenge U.S. naval dominance in the Western Pacific.

The concept of Anti-Access/Area Denial (A2/AD) is central to China's military strategy in the South China Sea. This approach involves deploying weapons systems and sensors that can detect, track, and potentially destroy enemy forces attempting to operate in the region. The goal is to create a zone of control that would make it extremely costly and risky for adversaries to intervene in a potential conflict.

The economic dimensions of China's South China Sea strategy are equally important. The region contains significant oil and natural gas reserves, estimated at 11 billion barrels of oil and 190 trillion cubic feet of natural gas. Control over these resources would enhance China's energy security and reduce its dependence on imports from potentially unstable regions.

Fishing rights represent another important economic dimension of the South China Sea disputes. The region's waters support some of the world's most productive fisheries, providing livelihoods for millions of people across multiple countries. China's use of large fishing fleets as tools of territorial assertion has created tensions with other claimant states and raised concerns about environmental sustainability.

The United States has responded to China's South China Sea activities through a combination of diplomatic, economic, and military measures. Freedom of Navigation Operations (FONOPs) involve U.S. naval vessels transiting through areas claimed by China to demonstrate that these claims are not recognized under international law. These operations have become more frequent and visible as tensions have increased.

The strengthening of alliance relationships has been a key component of the U.S. response to China's rise. The Quad partnership between the United States, Japan, Australia, and India has been revitalised as a mechanism for coordinating responses to Chinese assertiveness. The AUKUS partnership between Australia, the United Kingdom, and the United States represents an even deeper level of security cooperation focused specifically on the Indo-Pacific region.

Taiwan represents the most dangerous potential flashpoint in U.S.-China relations, with implications that extend far beyond the South China Sea. China considers Taiwan to be a renegade province that must eventually be reunified with the mainland, while the United States maintains a policy of strategic ambiguity regarding its commitment to Taiwan's defense. The potential for conflict over Taiwan has increased as China's military capabilities have grown and its rhetoric has become more assertive.

The semiconductor industry has become a critical dimension of U.S.-China competition, with Taiwan playing a central role as the world's leading producer of advanced computer chips. Taiwan Semiconductor Manufacturing Company (TSMC) produces the majority of the world's most advanced semiconductors, making the island's security a matter of global economic importance.

Economic interdependence between China and its neighbors creates complex dynamics that both constrain and enable conflict. Many countries in the region depend heavily on trade with China, making them reluctant to take strong positions on territorial disputes. However, this same economic relationship also gives China leverage that it can use to influence other countries' behavior.

The environmental consequences of activities in the South China Sea have become increasingly severe. China's island-building activities have destroyed coral reefs and marine ecosystems, while overfishing and pollution have degraded the marine environment. Climate change is also affecting the region, with rising sea levels potentially submerging some of the features that are the subject of territorial disputes.

The role of international law in the South China Sea disputes remains contentious. While the 2016 arbitration ruling clearly rejected China's claims, the lack of enforcement mechanisms for international legal decisions means that the ruling has had limited practical impact. This highlights broader questions about the effectiveness of international law in constraining great power behavior.

Regional countries have attempted to manage tensions through diplomatic mechanisms such as the Association of Southeast Asian Nations (ASEAN) and the Declaration of Conduct in the South China Sea. However, these efforts have been hampered by China's preference for bilateral negotiations and its resistance to multilateral approaches that might constrain its freedom of action.

The information warfare dimension of South China Sea competition involves efforts by all parties to shape international opinion and justify their positions. China has invested heavily in media and academic institutions to promote its narrative, while the United States and its allies have sought to highlight the importance of international law and freedom of navigation.

Looking forward, the South China Sea is likely to remain a central focus of great power competition for the foreseeable future. The resolution of territorial disputes will require either a negotiated settlement that addresses the legitimate interests of all parties or a shift in the regional balance of power that makes continued competition unsustainable.

The implications of South China Sea developments extend far beyond the region itself. The precedents set in this area will influence how territorial disputes are resolved in other parts of the world and will shape the broader evolution of the international system. The ability of the United States and China to manage their competition in the South China Sea without triggering a catastrophic conflict will be one of the defining challenges of the 21st century.

Understanding China's long game in the South China Sea is essential for comprehending the broader dynamics of great power competition and the challenges facing the international system. The decisions made by leaders in Beijing, Washington, and regional capitals will determine whether this vital region becomes a source of stability and prosperity or a trigger for conflict that could reshape the global order.

Chapter 11: Russia, Ukraine, and the European Powder Keg

The re-invasion of Ukraine by Russia in February 2022 shattered the illusion of a lasting peace in Europe and plunged the continent into its most dangerous security crisis since the end of the Cold War. This conflict is not merely a territorial dispute between two neighboring countries; it represents a fundamental challenge to the post-Cold War European security order, with profound implications for the future of NATO, the European Union, and the global balance of power.

The Return of Great Power Conflict to Europe

For decades, Europe has operated on the assumption that large-scale conventional warfare between major powers was a relic of the past. The continent's security architecture was built on a foundation of international law, arms control treaties, and cooperative security organisations designed to prevent the kind of catastrophic conflicts that defined the first half of the 20th century. Russia's full-scale aggression against Ukraine has demonstrated that this assumption was tragically misplaced.

The conflict has brought the brutal reality of industrial-scale warfare back to European soil, with trench warfare, artillery duels, and drone attacks reminiscent of the First and Second World Wars, but with the added dimension of 21st-century technology. The scale of the fighting has been immense, with hundreds of thousands of casualties on both sides and entire cities reduced to rubble. The war has also created the largest refugee crisis in Europe since the Second World

War, with millions of Ukrainians forced to flee their homes.

NATO's Renewed Purpose and Expansion

Russia's re-invasion of Ukraine has breathed new life into the North Atlantic Treaty Organisation (NATO), which had been struggling with questions about its purpose and relevance in the post-Cold War era. The threat of Russian aggression has galvanised the alliance, leading to a significant increase in defense spending by member states and the deployment of additional military forces to Eastern Europe to deter further Russian expansionism.

The most significant geopolitical consequence of the war has been the expansion of NATO to include Finland and Sweden, two countries that had long maintained policies of military neutrality. Their decision to join the alliance represents a fundamental shift in the European security landscape, extending NATO's border with Russia by more than 1,300 kilometers and transforming the Baltic Sea into what some analysts have called a "NATO lake." This expansion has significantly strengthened the alliance's position in Northern Europe but has also increased the potential for direct confrontation between NATO and Russian forces.

The European Union as a Geopolitical Actor

The war in Ukraine has also forced the European Union to assume a more significant role as a geopolitical actor. The EU has imposed unprecedented economic sanctions on Russia, provided billions of euros in financial and military assistance to Ukraine, and granted Ukraine candidate status for EU membership. These actions represent a significant departure from the EU's traditional focus on economic and regulatory issues and demonstrate a newfound willingness to use its collective power to address security challenges.

However, the war has also exposed the limitations of the EU's ability to act decisively in foreign and security policy. The requirement for unanimity among member states on key decisions has often led to delays and compromises, while differences in national interests and threat perceptions have created challenges for a unified response. The EU's dependence on the United States for its security remains a significant vulnerability, and the uncertainty from a second Trump presidency has raised concerns about the future of the transatlantic security relationship.

The Risk of Escalation and Nuclear Blackmail

The war in Ukraine has been characterised by a constant risk of escalation, with the potential for the conflict to spill over into neighboring countries or to escalate into a direct confrontation between Russia and NATO. Russian President Vladimir Putin has repeatedly engaged in nuclear blackmail, threatening to use nuclear weapons to deter Western support for Ukraine and to force a negotiated settlement on his terms. These threats have raised fears of a nuclear conflict in Europe for the first time in decades and have forced Western leaders to navigate a delicate balance between supporting Ukraine and avoiding a direct military confrontation with a nuclear-armed adversary.

The potential for miscalculation or accident to trigger a wider conflict remains a significant concern. The presence of large numbers of military forces in close proximity, the use of advanced weapons systems, and the constant risk of cyber-attacks and disinformation campaigns create a volatile environment in which a minor incident could quickly spiral out of control.

The Broader European Powder Keg

The war in Ukraine has also exacerbated existing tensions and created new fault lines across the European continent, turning the region into a volatile powder keg. The Western Balkans, in particular, has emerged as a key area of concern, with rising ethnic tensions, political instability, and growing Russian and Chinese influence threatening to reignite old conflicts. The unresolved status of Kosovo, the political crisis in Bosnia and Herzegovina, and the ongoing dispute between Serbia and its neighbors all represent potential flashpoints that could be exploited by external actors seeking to destabilise the region.

The conflict has also highlighted the vulnerability of other countries in Russia's neighborhood, such as Moldova and Georgia, which have their own unresolved territorial disputes with Russia and are facing increasing pressure from Moscow. The security of these countries is a key concern for NATO and the EU, as any further Russian aggression would have significant implications for European security.

The Future of European Security

The war in Ukraine has fundamentally altered the European security landscape, and the continent is now facing a new era of confrontation and competition with Russia. The decisions made in the coming months and years will shape the future of European security for generations to come. Key questions remain about the long-term strategy for dealing with a revanchist Russia, the future of the transatlantic security relationship, and the role of the European Union as a security actor.

The path to a stable and secure Europe will require a sustained commitment to collective defense, a significant increase in defense investment, and a renewed focus on building resilience to a wide range of threats, from conventional military aggression to cyber-attacks and disinformation. It will also require a clear-eyed understanding of the challenges ahead and a willingness to make difficult choices in the face of uncertainty. The European powder keg has been ignited, and the continent is now facing the difficult task of preventing the fire from spreading.

Part IV: The Information War

Chapter 12: The Algorithm Trap and the Erosion of Truth

In the digital age, the battle for truth has become one of the most critical fronts in the war that has already started. This is not a war fought with conventional weapons, but with algorithms, artificial intelligence, and sophisticated manipulation techniques that can reshape reality itself. The erosion of shared truth represents perhaps the greatest threat to democratic societies, as it undermines the very foundation upon which informed decision-making and democratic governance depend.

The Architecture of Algorithmic Manipulation

Social media algorithms have become the invisible architects of our information environment, determining what billions of people see, hear, and ultimately believe. These algorithms are not neutral arbiters of information; they are designed with specific objectives that often prioritise engagement over accuracy, emotion over reason, and controversy over consensus. The result is a digital ecosystem that systematically amplifies the most divisive and emotionally charged content while marginalising nuanced, factual reporting.

The mechanics of algorithmic manipulation are both sophisticated and insidious. Machine learning systems analyse vast amounts of user data to identify psychological triggers that generate engagement—clicks, shares, comments, and time spent on platform. Content that provokes strong emotional responses, particularly anger and outrage, is algorithmically rewarded with greater distribution. This creates a feedback loop where the most inflammatory and polarising content receives the widest reach, while measured, factual information struggles to compete for attention.

Research has demonstrated that false information spreads six times faster than true information on social media platforms, and this disparity is not accidental—it is a direct consequence of how these algorithms are designed to function. The pursuit of engagement metrics has created an information environment where truth is not just competing with falsehood on equal terms; it is systematically disadvantaged by the very systems that govern information distribution.

The Filter Bubble Phenomenon

One of the most pernicious effects of algorithmic curation is the creation of filter bubbles—personalised information environments that isolate users from diverse perspectives and conflicting information. These digital echo chambers are constructed through sophisticated profiling techniques that track user behavior, preferences, and social connections to create increasingly narrow information diets tailored to confirm existing beliefs and biases.

The filter bubble effect is particularly dangerous because it operates below the threshold of conscious awareness. Users are not typically informed about how their information environment is being curated, nor are they given meaningful choices about the algorithmic processes that determine what they see. This creates an illusion of choice and diversity while actually constraining the range of information and perspectives to which individuals are exposed.

The psychological impact of filter bubbles extends far beyond simple confirmation bias. When people are consistently exposed only to information that confirms their existing beliefs, they develop what researchers call "false consensus" to the mistaken belief that their views are more widely shared than they actually are. This phenomenon contributes to political polarisation and makes it increasingly difficult for people with different perspectives to find common ground or engage in productive dialogue.

The Weaponisation of Artificial Intelligence

The emergence of generative artificial intelligence has introduced new dimensions to information warfare that were previously confined to the realm of science fiction. Deepfake technology can now create convincing audio and video content featuring public figures saying or doing things they never actually said or did. While the most sophisticated deepfakes still require significant technical expertise to create, the barrier to entry is rapidly decreasing as AI tools become more accessible and user-friendly.

The threat posed by deepfakes extends beyond their potential to deceive audiences directly. Perhaps more significantly, the mere existence of deepfake technology creates what researchers call the "liar's dividend"—the ability for bad actors to dismiss authentic evidence as potentially fabricated. When any audio or video recording can potentially be dismissed as a deepfake, the evidentiary value of such recordings is fundamentally undermined, creating new opportunities for those seeking to escape accountability for their actions.

Artificial intelligence is also being used to generate vast quantities of synthetic text content that can flood information environments with false or misleading narratives. These AI-generated articles, social media posts, and comments can be produced at scale and distributed through networks of fake accounts, creating the appearance of grassroots support for particular viewpoints or the illusion of widespread belief in false information.

The Erosion of Institutional Trust

The systematic manipulation of information environments has contributed to a broader erosion of trust in traditional institutions, including journalism, science, and democratic governance. When people are constantly exposed to conflicting information and struggle to distinguish between reliable and unreliable sources, they often retreat into tribal loyalties and ideological certainties that provide psychological comfort but undermine their ability to engage with complex realities.

This erosion of institutional trust creates a vicious cycle. As people lose faith in traditional sources of authority and expertise, they become more susceptible to alternative narratives that may be less rigorous or entirely fabricated. This, in turn, further undermines the credibility of legitimate institutions and creates space for bad actors to exploit public confusion and uncertainty.

The phenomenon of "truth decay"—the diminishing reliance on facts and analysis in public life—has become a defining characteristic of contemporary political discourse. When shared standards of evidence and reasoning break down, democratic deliberation becomes increasingly difficult, and societies become vulnerable to manipulation by those willing to exploit the resulting confusion.

The Global Information War

The manipulation of information environments is not merely a byproduct of technological development; it has become a deliberate strategy employed by state and non-state actors seeking to advance their interests through the destabilisation of their adversaries. Foreign interference in democratic processes increasingly relies on information warfare techniques designed to exacerbate existing social divisions and undermine confidence in democratic institutions.

The sophistication of these operations has evolved dramatically in recent years. Rather than simply promoting particular candidates or policies, modern information warfare campaigns often focus on amplifying existing social tensions and promoting general cynicism about the possibility of truth itself. By fostering an environment where all information is viewed with suspicion, these campaigns make it easier for authoritarian actors to dismiss legitimate criticism and accountability efforts.

The global nature of digital platforms means that information warfare campaigns can have effects far beyond their intended targets.

Disinformation created to influence one country's elections can spread across borders and languages, creating cascading effects that destabilise multiple societies simultaneously. This interconnectedness makes it increasingly difficult to contain the effects of information warfare and requires unprecedented levels of international cooperation to address effectively.

The Psychology of Digital Manipulation

Understanding the psychological mechanisms that make people vulnerable to digital manipulation is crucial for developing effective countermeasures. Human beings have evolved cognitive shortcuts and biases that served them well in small-scale social environments but can be exploited in the context of digital information systems. Confirmation bias, availability probing, and social proof are just a few of the psychological tendencies that can be weaponised through algorithmic manipulation.

The addictive design of social media platforms compounds these vulnerabilities by creating psychological dependencies that make users more susceptible to manipulation. Features like infinite scroll, variable reward schedules, and social validation mechanisms are deliberately designed to maximise user engagement, often at the expense of user wellbeing and critical thinking capacity.

The speed and volume of information in digital environments also overwhelm human cognitive capacity, making it difficult for people to carefully evaluate the credibility and accuracy of the information they encounter. This cognitive overload creates opportunities for bad actors to slip false or misleading information past people's critical defenses, particularly when such information is emotionally compelling or confirms existing beliefs.

The Challenge of Content Moderation

Social media platforms face enormous challenges in moderating content at the scale required by their global user bases. The sheer volume of content uploaded every minute makes comprehensive human review impossible, while automated systems struggle with the nuance and context required to make accurate judgements about complex information. This creates opportunities for sophisticated actors to game moderation systems and evade detection while spreading false or harmful content.

The global nature of these platforms also creates complex jurisdictional challenges. Content that is legal and acceptable in one country may violate laws or social norms in another, forcing platforms to make difficult decisions about how to balance free expression with harm prevention across diverse cultural and legal contexts.

The privatisation of content moderation decisions also raises important questions about democratic accountability. When a small number of technology companies have the power to determine what information billions of people can access, this represents a significant concentration of power that may not be subject to adequate democratic oversight or accountability mechanisms.

Building Resilience Against Information Warfare

Addressing the challenges posed by algorithmic manipulation and information warfare requires a multi-faceted approach that combines technological solutions, regulatory frameworks, and educational initiatives. Digital literacy education must become a priority for educational systems, helping people develop the skills needed to navigate complex information environments and identify potential manipulation attempts.

Technological solutions, including improved detection systems for synthetic media and more transparent algorithmic processes, can help level the playing field between those seeking to manipulate information and those working to preserve its integrity. However, technological solutions alone are insufficient; they must be combined with broader efforts to strengthen democratic institutions and rebuild social trust.

Regulatory frameworks must evolve to address the unique challenges posed by digital information systems while preserving fundamental rights to free expression and privacy. This requires careful balance and ongoing adaptation as technologies and tactics continue to evolve.

Perhaps most importantly, societies must work to rebuild shared standards of evidence and reasoning that can serve as a foundation for democratic deliberation. This requires not just technical solutions but a renewed commitment to the values and practices that make democratic governance possible in an age of information abundance and manipulation.

The algorithm trap represents one of the most sophisticated and dangerous weapons in the arsenal of those seeking to undermine democratic societies. Understanding how these systems work and developing effective countermeasures is essential for preserving the possibility of truth and democratic governance in the digital age. The stakes could not be higher: the future of democracy itself may depend on our ability to navigate the complex challenges posed by the erosion of truth in our interconnected world.

Part V: What Comes Next

Chapter 13: The Psychology of Leaders in Crisis

The decisions made by political leaders during times of crisis can determine the fate of nations and the lives of millions of people. Understanding the psychological factors that influence leadership decision-making in high-stress situations is crucial for comprehending how conflicts escalate, how crises are resolved, and how catastrophic mistakes can be avoided. The psychology of crisis leadership reveals both the remarkable capacity of human beings to rise to extraordinary challenges and their vulnerability to cognitive biases and emotional pressures that can lead to disastrous outcomes.

The Cognitive Burden of Crisis Leadership

Political leaders facing existential threats operate under psychological pressures that few human beings ever experience. The weight of responsibility for millions of lives, combined with incomplete information, time constraints, and the knowledge that their decisions will be scrutinised by history, creates a cognitive burden that can overwhelm even the most experienced leaders.

Research in cognitive psychology has identified several factors that can impair decision-making under extreme stress. Time pressure reduces the ability to process information thoroughly and consider alternative options. Information overload can lead to cognitive paralysis or reliance on simplified probing that may not be appropriate for complex situations. The stress response itself can narrow attention and reduce creative problem-solving capacity.

The phenomenon of "groupthink" becomes particularly dangerous in crisis situations, where the pressure for consensus and the need for rapid decision-making can suppress dissenting voices and critical analysis.

Leaders surrounded by advisors who share similar backgrounds and perspectives may develop false confidence in their decisions while remaining blind to alternative viewpoints or potential risks.

Historical Patterns in Crisis Decision-Making

Historical analysis of major crises reveals recurring patterns in how leaders respond to extreme pressure. The Cuban Missile Crisis of 1962 provides a classic example of how psychological factors can both escalate and de-escalate dangerous situations. President John F. Kennedy's initial inclination toward military action was tempered by advisors who encouraged careful deliberation and consideration of Soviet perspectives. The eventual resolution required both leaders to find face-saving compromises that allowed them to step back from the brink of nuclear war.

Conversely, the outbreak of World War I demonstrates how psychological factors can contribute to catastrophic escalation. The mobilization schedules and alliance commitments that European leaders had created in peacetime became psychological traps during the July Crisis of 1914. Once the crisis began, leaders felt compelled to honor their commitments and maintain their credibility, even as they recognised that their actions were leading toward a devastating war that none of them actually wanted.

The role of personal relationships between leaders can be crucial in crisis situations. The personal animosity between Adolf Hitler and Joseph Stalin contributed to the breakdown of their non-aggression pact and Germany's disastrous invasion of the Soviet Union. Conversely, the personal rapport between Ronald Reagan and Mikhail Gorbachev helped facilitate the end of the Cold War despite significant ideological differences between their countries.

The Impact of Personality on Crisis Leadership

Individual personality traits can have enormous consequences when leaders face existential threats. Narcissistic leaders may be unable to acknowledge mistakes or accept advice that contradicts their self-image. Paranoid leaders may interpret neutral actions as hostile and respond with disproportionate force. Impulsive leaders may make irreversible decisions without adequate consideration of consequences.

The concept of "malignant narcissism" has been particularly relevant in understanding how certain types of leaders can pose existential threats to their own societies and the international system. Leaders who combine narcissistic personality traits with paranoid thinking, a willingness to use violence, and a lack of empathy for others can make decisions that prioritise their personal survival over the welfare of their people or global stability.

However, crisis situations can also reveal positive leadership qualities that might not be apparent during normal times. The ability to remain calm under pressure, to inspire confidence in others, and to make difficult decisions based on principle rather than political expediency can emerge during moments of extreme challenge.

Winston Churchill's leadership during World War II exemplifies how personal courage and rhetorical skill can rally a nation during its darkest hours.

The Role of Advisors and Institutional Constraints

The quality of advice that leaders receive during crises can be as important as their own decision-making capabilities. Effective advisory systems provide leaders with diverse perspectives, challenge their assumptions, and help them consider the long-term consequences of their actions. The structure of decision-making processes can either facilitate or hinder good judgement during crisis situations.

Institutional constraints can serve as important safeguards against impulsive or destructive decision-making. Constitutional systems that require consultation with legislative bodies, military structures that provide checks on civilian authority, and international institutions that create forums for dialogue can all help prevent crisis escalation. However, these same constraints can also be seen as obstacles by leaders who believe that rapid, decisive action is necessary.

The role of intelligence services in providing information to leaders during crises is particularly critical. Intelligence failures can lead to catastrophic miscalculations, while accurate intelligence can provide the foundation for effective crisis management. However, the interpretation of intelligence is often influenced by the preconceptions and biases of both intelligence analysts and political leaders.

Cultural and Historical Influences

The cultural background and historical experience of leaders significantly influence how they interpret and respond to crisis situations. Leaders from countries with histories of invasion or occupation may be more sensitive to perceived threats and more willing to use preemptive force. Leaders from stable, prosperous societies may underestimate the willingness of others to accept enormous costs to achieve their objectives.

Historical analogies play a powerful role in crisis decision-making, as leaders often interpret current situations through the lens of past events. The "Munich analogy"—the belief that appeasement of aggressive dictators only encourages further aggression—has influenced numerous decisions by Western leaders since World War II. However, the inappropriate application of historical analogies can lead to misguided policies that escalate rather than resolve crises.

Generational differences can also be significant, as leaders who lived through major historical events may have different risk tolerances and strategic perspectives than those who know these events only through study. The passing of the generation that experienced World War II and the Cold War firsthand may change how future leaders approach international crises.

The Psychology of Nuclear Decision-Making

The unique psychological pressures associated with nuclear weapons create special challenges for crisis leadership. The knowledge that a single decision could result in the deaths of millions of people and the destruction of civilisation itself creates psychological burdens that no human being is truly prepared to bear. The compressed timelines associated with nuclear threats leave little time for careful deliberation or consultation.

The concept of "nuclear taboo"—the strong psychological and moral inhibition against using nuclear weapons—has been an important factor in preventing nuclear conflict since 1945. However, this taboo may be weaker among leaders who did not live through the Cold War or who come from different cultural backgrounds. The proliferation of nuclear weapons to additional countries increases the risk that leaders without strong inhibitions against nuclear use may gain access to these weapons.

The psychological effects of nuclear threats on decision-making can be paradoxical. While the enormous destructive potential of nuclear weapons should encourage caution, the fear of appearing weak or irresolute may push leaders toward more aggressive postures. The need to maintain credible deterrence requires leaders to convince adversaries that they would be willing to use nuclear weapons under certain circumstances, even if they privately hope never to face such a decision.

Modern Challenges to Crisis Leadership

Contemporary leaders face psychological challenges that their predecessors did not encounter. The speed of modern communications means that crises can escalate much more rapidly than in the past, leaving less time for careful deliberation. The constant scrutiny of social media and 24-hour news cycles creates additional pressure to appear decisive and in control.

The complexity of modern threats, from cyber-attacks to climate change to pandemic diseases, requires leaders to understand and respond to challenges that may be outside their areas of expertise. The interconnectedness of global systems means that local crises can quickly have global implications, requiring leaders to consider consequences far beyond their immediate responsibilities.

The erosion of trust in institutions and expertise makes it more difficult for leaders to build consensus for difficult decisions. When significant portions of the population doubt the credibility of government institutions, scientific expertise, or media reporting, leaders may find it challenging to communicate effectively about crisis situations or build support for necessary but unpopular measures.

Implications for Crisis Prevention and Management

Understanding the psychology of crisis leadership has important implications for how societies can better prepare for and manage existential threats. Training programs for political leaders should include education about cognitive biases, stress management techniques, and historical case studies of crisis decision-making. Advisory systems should be designed to provide diverse perspectives and challenge leader assumptions.

International institutions and diplomatic protocols should be strengthened to provide forums for communication during crises and to create face-saving opportunities for leaders to step back from dangerous confrontations. Early warning systems and confidence-building measures can help prevent misunderstandings that might escalate into serious crises.

Perhaps most importantly, societies should work to select and support leaders who demonstrate the psychological qualities necessary for effective crisis management: emotional stability, intellectual humility, the ability to listen to diverse viewpoints, and the moral courage to make difficult decisions based on principle rather than political expediency.

The psychology of crisis leadership reveals both the enormous potential and the dangerous limitations of human decision-making under extreme pressure. By understanding these psychological factors, societies can better prepare for the crises that will inevitably arise and increase the likelihood that their leaders will make decisions that preserve rather than destroy the foundations of human civilisation.

Epilogue: A Path Forward and the Conversation Continues

As we reach the end of this examination of the war that has already started, it becomes clear that we stand at a critical juncture in human history. The threats we have explored—from cyber warfare and nuclear proliferation to information manipulation and great power competition—are not distant possibilities but present realities that shape our world every day. The question is not whether these challenges exist, but how we will choose to respond to them.

The conversation that inspired this book began with a simple question about whether we are already at war. Through our journey together, we have discovered that the answer is unequivocally yes—but this war looks nothing like the conflicts of the past. It is fought in cyberspace and financial markets, through algorithms and artificial intelligence, in the minds of ordinary citizens who may not even realise they are participants in a global struggle for the future of human civilisation.

This new form of warfare presents both unprecedented dangers and unique opportunities. The same technologies that enable devastating cyber-attacks also provide tools for education, communication, and cooperation. The global interconnectedness that creates vulnerabilities also creates possibilities for collective action and shared solutions. The information systems that can be weaponised for manipulation can also be harnessed for transparency and accountability.

The path forward requires us to abandon comfortable illusions about the nature of contemporary conflict while maintaining hope about our capacity to shape a better future. We must acknowledge the reality of the threats we face without succumbing to despair or paralysis. We must prepare for the worst while working toward the best possible outcomes.

Individual action remains crucial in this new landscape of conflict. Every person who develops digital literacy skills, who questions information sources, who builds genuine relationships across political and cultural divides, who supports democratic institutions and international cooperation, contributes to our collective resilience. The war that has already started is not just fought by governments and militaries—it requires the active participation of informed and engaged citizens.

The choices we make in the coming years will determine whether the 21st century becomes an era of unprecedented human flourishing or a period of conflict and decline. The technologies that threaten us also offer solutions to humanity's greatest challenges, from climate change to poverty to disease. The global connections that create vulnerabilities also enable cooperation on a scale never before possible in human history.

The conversation that began with a podcast episode about global threats must continue in classrooms and boardrooms, in legislative chambers and diplomatic conferences, in families and communities around the world. We must talk honestly about the dangers we face while working together to build the institutions, relationships, and capabilities needed to address them effectively.

The war that has already started is not inevitable in its outcomes.

Human beings have repeatedly demonstrated their capacity to overcome seemingly insurmountable challenges through courage, creativity, and cooperation.

The same species that created nuclear weapons also created international institutions to control them. The same technologies that enable cyber warfare also enable global communication and collaboration.

Our task is to ensure that the better angels of human nature prevail in this new era of conflict and competition. This requires not just technical solutions but moral leadership, not just individual preparation but collective action, not just national responses but international cooperation. The stakes could not be higher, but neither could the potential rewards of success.

The conversation continues, and your voice matters. The war that has already started will be won or lost not just by governments and militaries, but by ordinary people who choose to engage with these challenges rather than ignore them, who seek truth rather than comfort, who build bridges rather than walls. The future remains unwritten, and we all have a role to play in determining how the story ends.

The path forward is neither simple nor certain, but it is possible. By understanding the nature of contemporary conflict, preparing for its challenges, and working together toward common solutions, we can navigate the dangers ahead and build a more secure and prosperous world for future generations. The war that has already started need not define our destiny—if we have the wisdom to understand it and the courage to act.

Summary

This book is a gateway to a new movement, a new way of looking at life (and death), and to explain why we all need to understand what is going on in the world right now and what we have to do to stop this from destroying civilization.

I am not trying to use this book to scare people, raise publicity or get rich quick through manipulation of media and minds, believe me that is already happening, and I am just pointing this out to you. In fact, I am offering real world options and solutions of how we can all work together to improve our lives and opportunities.

By taking responsibility for how we digest and react to news stories and how we either knowingly or unknowingly become part of the underlying issue and spread the news and hate that is behind these narratives. Narratives that are being deliberately spread to cause us all to disengage with uncomfortable but necessary debates and thoughts provoking civil conversations about some of the key issues we are facing right now.

These issues will vary for everyone depending on your geographical area you live in, but nevertheless it is important that you stop and take a breath and make informed decisions about what to do next.

Some of the issues that I am seeing are very clear, and I'm sure once this book is published and people read and understand the messages within, there will be hundreds of similar examples that you can clearly point out in your region of the world.

A recent report has flagged major platforms like Facebook and X (formerly Twitter) as "vectors of hate", citing their role in amplifying extremist content through algorithms and bots. This has reignited debates around regulation and platform accountability.

I have listed some of these current tensions as key examples:

Religious & Ethnic Tensions

France: Ongoing tensions around secularism and Islam have led to protests and attacks on mosques, especially following controversial legislation banning religious symbols in public spaces.

India: Communal violence between Hindu and Muslim communities has flared in several regions, often triggered by misinformation spread via social media.

Germany and Australia: Synagogues and Jewish institutions and even restaurants have faced vandalism and threats, particularly amid rising far- right sentiment and backlash over Middle East conflicts.

Corporate Symbolism & Political Backlash

United States (Tesla): Tesla dealerships and charging stations have been targeted with arson and vandalism following Elon Musk's political involvement in the Trump administration.

Anti-Immigration & Nationalist Violence

Northern Ireland: Riots erupted in Ballymena after two Romanian teens were charged with sexual assault. The unrest quickly escalated into anti-immigrant violence, with homes and businesses targeted, and racist slogans like "Veterans before refugees" displayed on bonfires.

Greece & Italy: Migrant camps have been attacked, and far-right groups have staged violent protests against refugee resettlement.

Sweden: Quran burnings and anti-immigrant demonstrations have sparked international outrage and domestic unrest.

Social Media Censorship & Shutdowns

India & Myanmar: Topped the list for internet shutdowns in 2024, often timed around protests or exams. India alone had 84 shutdowns, disrupting communication and sparking outrage.

Brazil & Turkey: Temporarily banned platforms like X (formerly Twitter) and Instagram during political unrest, citing misinformation concerns.

Russia & China: Continued aggressive censorship, blocking Western platforms and isolating their digital ecosystems.

AI-Driven Misinformation & Deepfakes

USA (Maryland): A high school principal was falsely accused via a viral deepfake audio clip. The scandal exposed how AI-generated content can incite racial and antisemitic outrage.

Israel–Palestine Conflict: Meta faced backlash for allegedly shadow banning pro-Palestinian content. Human Rights Watch documented over 1,050 suppressions on Instagram and Facebook.

Political Manipulation & Disinformation

Global Trends: Over 81 countries were found to be running organised social media manipulation campaigns, often using bots and paid influencers to sway public opinion.

Meta's Fact-Checking Rollback: Meta ended its third-party fact-checking program, sparking fears that misinformation will spread unchecked. Critics say this favors engagement over truth.

These are happening right now and are all in free democratic nations that you wouldn't necessarily expect to see such tensions rising…so why is this happening?

This is what my book has tried to relay to provide an overview of the bad actors and nations, generally the non-free democratic nations, that have the most to gain from separation and the division and discord of people in these free countries.

This is not to say that there are real issues that need resolving, there clearly are. Mass immigration from the Middle East and Africa to Europe, Great Britain and Ireland is clearly having a major impact on these economies and people. However, as free democratic nations there are better ways to rally together to bring change through political action, not through violence and rioting of your own countries, towns and cities where the local women and children are being caught up these hate driven and media induced riots.

Unlike the non-free democratic nations that cannot assemble peacefully to raise issues due to the real fear of arrest, beatings or worse. We are not left without peaceful options to highlight group concerns and ensure that our voices are being heard, and positive actions are put in place to make change. If this does mean a change of political parties or governments, then so be it, but let's do this together not tear each other apart trying to get the same changes.

As I have noted throughout my book, these are real issues happening right now and you all need to at least be aware of what is going on and what YOU can do about it. Yes, YOU can make a big difference to what is happening and the trajectory going forward. Some of the simple things that you can start with right now I have summarised for your consideration:

Personal & Societal Resilience Strategies

Critical Thinking & Media Literacy: Individuals need to sharpen their ability to discern truth from misinformation, especially in a landscape dominated by algorithmic manipulation.

Disciplined Information Consumption: Develop intentional media habits to avoid falling into polarised echo chambers and fear-based narratives.

Community Engagement: Building strong local networks was highlighted as a buffer against societal fragmentation and a way to foster collective problem-solving.

Geopolitical Adaptation & Safety Planning

Diplomatic Hope: Maintaining faith in diplomacy and international cooperation as tools to de-escalate tensions.

Navigating Emerging Technologies

AI Awareness & Preparedness: Be aware of deepfakes and autonomous weapons use and encourage proactive education and ethical development of AI systems.

Cybersecurity Literacy: Understanding digital vulnerabilities as essential for both individuals and institutions to protect against manipulation and sabotage.

Policy & Institutional Opportunities

Reforming Nuclear Policy: Encourage renewed global dialogue around tactical nuclear weapons to prevent their normalisation and potential use.

Revitalising Intelligence Oversight: Push for more transparent and accountable intelligence operations oversight to rebuild public trust.

Whilst these may seem to be unachievable in the short term together we can strive to not let these fall by the wayside and look to improve societies and the governments that are responsible to the people that have elected them. Together we can force them out of office through the people's combined power if they fail to listen.

Here are some simple steps that we can all adopt to stay Mentally Clear & Media-Savvy:

Limit Screen Time: Schedule "offline hours" daily — especially in the evening — to reduce emotional fatigue and regain focus.

Curate Media Sources: Follow a handful of reputable, fact-checked outlets with diverse perspectives. Avoid those relying on outrage or clickbait.

Fact-Check Routinely: Use tools like Snopes, Media Bias/Fact Check, and government fact-checkers before sharing headlines or viral content.

Recognise Psychological Triggers: Learn how outrage, fear, and tribal loyalty are used to manipulate attention. Pause before reacting emotionally.

Reconnect with Real People & Purpose

Join Like-Minded Clubs: Whether it's hiking, philosophy, gardening, or writing — these groups foster authentic connection and shared values.

Volunteer Locally: Help a community garden, shelter, or literacy program. It's a powerful antidote to cynicism and division.

Host Media-Free Social Events: Create phone-free dinner nights, book circles, or open discussions with trusted friends via any hobbies, sports or similar activities.

Recharge in Nature & the Present Moment

Daily Nature Breaks: Even short walks in green spaces calm the nervous system and shift mental gears from reactive to reflective.

Practice Mindfulness or Meditation: Just 10 minutes of breathing or quiet can increase resilience against media noise.

Create Rituals of Stillness: Journal at sunrise, sip tea without screens, or stargaze before bed — these moments ground awareness.

Foster Kindness & Curiosity

Assume Positive Intent: When engaging with others online or offline, begin with empathy — even if their views differ.

Listen More Than You Argue: It's tempting to debate, but deep listening disarms hostility and reveals common ground.

Model Grace: Be the calm in the storm. How you respond to chaos shapes the environment more than any headline.

Dedication

This book was greatly inspired by the Diary of a CEO (DOAC) podcast: WW3 Threat Assessment: The West Is Collapsing, Can We Stop It?! They Want You Confused & Obedient!

This 2025 DOAC roundtable podcast brings together 3 top experts: Former CIA intelligence officer Andrew Bustamante, nuclear war journalist Annie Jacobsen, and global politics expert Benjamin Radd to discuss the biggest threats facing the world right now and has over 3million views at the time of writing this.

DOAC is a raw, curiosity driven podcast hosted by Steven Bartlett, where he interviews influential thinkers, entrepreneurs, and cultural icons to uncover deep insights about success, failure, mental health, and personal growth. It's part self-help, part storytelling — offering unfiltered conversations that aim to challenge conventional wisdom and inspire meaningful change (I am not getting paid to promote DOAC if you're asking).

I am a regular podcast follower and researcher, and I carefully ensure that I try to get a balanced viewpoint from checking out various providers to hear different opinions.

Staying abreast of world news is not a bad thing at all, it is just that we need to be aware of exactly what is going on behind the scenes and not get caught going down rabbit holes that have been set up to trap us.

I am also going to show how important this is for us all and hope to soon be launching my own website as an umbrella movement, to facilitate the building of a more resilient and happier world, that provides the opportunity to meet like-minded people and put loneliness back in the box it has escaped from.

Global Loneliness Overview

1 in 6 people worldwide — nearly 1 billion individuals — experience loneliness regularly.

The WHO reports that loneliness contributes to over 871,000 deaths annually, with 100 people dying every hour from loneliness-related causes.

Young adults (19–29) report the highest rates of loneliness, with 27% feeling fairly or very lonely.

Older adults (65+) have lower rates, but still 17% report frequent loneliness.

Australia-Specific Trends

62% of young adults and 46% of seniors in Australia report feeling lonely. 55% of Australians say they lack companionship at least sometimes.

Health Impacts

Loneliness increases the risk of:

- Stroke by 32%

- Heart disease by 29%

- Dementia by 50% in older adults

- Its health impact is comparable to smoking 15 cigarettes a day.

Digital & Social Factors

Excessive screen time and online toxicity are major contributors, especially among youth.

Despite hyper-connectivity, many people feel emotionally disconnected and socially isolated.

Personally, I like to read a variety of interesting books, meditate (not very well yet but that's another story) and take walks in nature. These are free, when utilising your local library, and easy for anyone to do as a starting point to make positive change. Part of the new Website I am creating will cover book recommendations, but to start this off I can recommend a book that I have just started and am finding fascinating, insightful and thought provoking:

The Tibetan Book of Living and Dying (25th Anniversary Edition)
by Sogyal Rinpoche

Core Message & Purpose

This spiritual classic blends Tibetan Buddhist teachings with Western insights to guide readers through the mysteries of death, dying, and rebirth— but more importantly, it's a manual for how to live fully and compassionately. Rinpoche draws from the Bardo Thodol (Tibetan Book of the Dead) and his own experiences to demystify death and present it as a natural, transformative part of life.

Why Accepting Death Improves Life

Impermanence as a Teacher: By regularly contemplating death, we strip away illusions of permanence and control. This leads to greater clarity, humility, and appreciation for the present moment.

Freedom from Fear: Understanding death reduces anxiety and denial, allowing us to live with courage and authenticity.

Prioritising What Matters: Awareness of mortality helps us focus on meaningful relationships, spiritual growth, and inner peace — rather than material distractions.

Compassion & Connection: Accepting our shared fate fosters empathy. We become more patient, kind, and present with others.

Spiritual Readiness: Preparing for death through meditation and reflection cultivates a calm, spacious mind — which benefits both living and dying.

Practical Teachings

Meditation & Mind Training: Practices like Dzogchen, Tonglen, and Phowa help us connect with our true nature and prepare for death with grace.

Bardo Awareness: Life itself is a bardo — a transitional state. Recognising this helps us navigate change and suffering with wisdom.

Caring for the Dying: The book offers guidance on how to support others through death with love, presence, and spiritual care.

In essence, Rinpoche teaches that death is not the end, but a mirror — reflecting how we've lived and what we've valued. By facing it regularly, we awaken to life's sacredness and become more intentional, compassionate, and free.

I have other read books on Death before like SADHGURU's DEATH – AN INSIDE STORY, which I can also recommend. Whilst it seems like a depressing subject we all have to face this is the reality, no matter how much money or power you have, you find that you feel you have to live life better after reading these books.

In closing I would like to thank you for reading my first book and hope through my website, details to follow shortly, we can join people together and stop letting others try to tear us apart.

I also volunteer for the Red Cross Australia, so a bit shout out to all the volunteers there they do great work and prove that as a human we care about everyone else deep down. Especially when they're going through the worse disasters that unfortunately happening far too often due to climate change and the inaction of the largest emitters in the world.

The End…for now.

AI Assistance Disclosure

This book was developed with the support of artificial intelligence tools to assist in content generation, quality assurance, and editorial refinement. All final decisions regarding the manuscript's structure, tone, and messaging were made by the author to ensure alignment with the intended vision and values.

References and Source Credits

Primary Sources and Expert Interviews

Diary of a CEO Podcast. "WW3 Threat Assessment: The West Is Collapsing, Can We Stop It?! They Want You Confused & Obedient!" Host: Steven Bartlett. Guests: Andrew Bustamante (Former CIA Intelligence Officer), Annie Jacobsen (Nuclear War Journalist), Benjamin Radd (Global Politics Expert). 2025.

Chapter 1: The Conversation That Changed Everything

Bustamante, Andrew. Interview on Diary of a CEO Podcast, 2025. Jacobsen, Annie. Interview on Diary of a CEO Podcast, 2025.Radd, Benjamin. Interview on Diary of a CEO Podcast, 2025.

Chapter 2: Digital Warfare and the New Battleground

FireEye. "Highly Evasive Attacker Leverages SolarWinds Supply Chain to Compromise Multiple Global Victims With SUNBURST

Backdoor." December 13, 2020.

Cybersecurity and Infrastructure Security Agency. "Alert (AA21-008A): Malicious Cyber Activity Attributed to Russian Government Actors." January 8, 2021.

Colonial Pipeline Company. "Colonial Pipeline System Disruption Update." May 2021.

Mandiant. "M-Trends 2024: A View from the Front Lines." 2024.

Chapter 3: The Forgotten Wars - Sudan, Myanmar, and Haiti

Armed Conflict Location & Event Data Project (ACLED). "Sudan Conflict Observatory." 2024.

United Nations Office for the Coordination of Humanitarian Affairs. "Sudan Humanitarian Response Plan 2024." 2024.

Famine Review Committee. "Famine Determination for Darfur, Sudan." August 2024.

Myanmar Witness. "Myanmar Civil War Documentation Project." 2024.

Association of Southeast Asian Nations. "ASEAN Leaders' Statements on Myanmar." 2021-2024.

United Nations Security Council. "Resolution 2699: Multinational Security Support Mission to Haiti." October 2, 2023.

Chapter 4: The AI Revolution in Warfare

U.S. Department of Defense. "Summary of the 2018 Department of Defense Artificial Intelligence Strategy." 2018.

Campaign to Stop Killer Robots. "Country Positions on Lethal Autonomous Weapons Systems." 2024.

Convention on Certain Conventional Weapons. "Report of the 2023 Meeting of Experts on Lethal Autonomous Weapons Systems."

Chapter 5: Speed Kills: The Hypersonic Threat

Congressional Research Service. "Hypersonic Weapons: Background and Issues for Congress." Updated 2024.

Center for Strategic and International Studies. "Missile Defense Project: Hypersonic Weapons." 2024.

Chapter 6: Inside the Nuclear Launch Process

Jacobsen, Annie. "Nuclear War: A Scenario." Dutton, 2024.

Federation of American Scientists. "Nuclear Notebook Series." Bulletin of the Atomic Scientists, 2024.

Chapter 7: Surviving the Nuclear Threat

Federal Emergency Management Agency. "Nuclear Explosion Response Guidelines." 2022.

Centers for Disease Control and Prevention. "Radiation Emergency Preparedness and Response." 2024.

Chapter 8: Israel, America, and the Proxy Game

Congressional Research Service. "U.S. Foreign Aid to Israel." Updated 2024. Institute for the Study of War. "Iran's Axis of Resistance." 2024.

Abraham Accords Peace Institute. "Annual Report 2024." 2024.

Chapter 9: Economic Warfare in the 21st Century

Peterson Institute for International Economics. "Economic Sanctions Database." 2024.

Atlantic Council. "Economic Statecraft Series." 2024.

Chapter 10: China's Long Game and the South China Sea

Asia Maritime Transparency Initiative. "South China Sea Tracker." Center for Strategic and International Studies, 2024.

Permanent Court of Arbitration. "The South China Sea Arbitration (The Republic of the Philippines v. The People's Republic of China)." Case No. 2013-19, Award of 12 July 2016.

U.S. Department of Defense. "Military and Security Developments Involving the People's Republic of China 2024." Annual Report to Congress.

Chapter 11: Russia, Ukraine, and the European Powder Keg

Institute for the Study of War. "Russian Offensive Campaign Assessment." Daily updates, 2022-2024.

NATO. "Strategic Concept 2022." Adopted at the Madrid Summit, June 29, 2022.

European Council. "EU Response to Russia's Aggression Against Ukraine." Updated 2024.

Chapter 12: The Algorithm Trap and the Erosion of Truth

Vosoughi, Soroush, Deb Roy, and Sinan Aral. "The Spread of True and False News Online." Science 359, no. 6380 (2018): 1146-1151.

Reuters Institute for the Study of Journalism. "Digital News Report 2024." Oxford University.

Partnership on AI. "Synthetic Media Framework." 2024.

Chapter 13: The Psychology of Leaders in Crisis

Janis, Irving L. "Groupthink: Psychological Studies of Policy Decisions and Fiascoes." Houghton Mifflin, 1982.

Kahneman, Daniel. "Thinking, Fast and Slow." Farrar, Straus and Giroux, 2011.

Allison, Graham T. "Essence of Decision: Explaining the Cuban Missile Crisis." Little, Brown, 1971.

Additional Academic and Policy Sources

RAND Corporation. "Truth Decay: An Initial Exploration of the Diminishing Role of Facts and Analysis in American Public Life." 2018.

Brookings Institution. "Order from Chaos: Foreign Policy in a Troubled World." Various reports, 2024.

Council on Foreign Relations. "Global Conflict Tracker." 2024.

International Institute for Strategic Studies. "The Military Balance 2024." 2024.

Stockholm International Peace Research Institute. "SIPRI Yearbook 2024: Armaments, Disarmament and International Security." 2024.

Data Sources and Statistical References

World Bank. "World Development Indicators." 2024.

International Monetary Fund. "World Economic Outlook Database." October 2024.

Uppsala Conflict Data Program. "UCDP Conflict Encyclopedia." 2024.

Note on Sources

This book draws primarily from open-source intelligence, academic research, policy analysis, and expert commentary available in the public domain. While every effort has been made to verify information accuracy, the rapidly evolving nature of contemporary conflicts means that some details may have changed since publication. Readers are encouraged to consult current sources for the most up-to-date information on ongoing situations.

The author acknowledges that access to classified information or insider perspectives might provide additional insights not available through public sources. However, the analysis presented here is based on the best available open-source information and expert analysis accessible to informed citizens and researchers.

Disclaimer

The views and analysis presented in this book are those of the author and do not necessarily reflect the official positions of any government, organisation, or institution. The book is intended for educational and informational purposes and should not be considered as professional advice for policy-making or strategic planning.

Index

A

B

C

D

N

P

R

S

T

Printed in Dunstable, United Kingdom